T0184673

Laser Shocking Nano-Crystallization
and High-Temperature Modification Technology

Xudong Ren

Laser Shocking Nano-Crystallization and High-Temperature Modification Technology

 Springer

Xudong Ren
Jiangsu University
Zhenjiang
People's Republic of China

ISBN 978-3-662-51502-0   ISBN 978-3-662-46444-1   (eBook)
DOI 10.1007/978-3-662-46444-1

Springer Heidelberg New York Dordrecht London

Printed on acid-free paper

Springer-Verlag GmbH Berlin Heidelberg is part of Springer Science+Business Media
(www.springer.com)

Preface

Laser shock processing, or laser shock peening (LSP), is the process of hardening or strengthening metal using a powerful laser. It could generate a layer of residual compressive stress on a surface of metallic materials and alloys that is several times deeper than that attainable from conventional shot peening treatments (shot peening), which has been successfully applied to improve fatigue performance of metallic components.

LSP is a new surface modification technology, which uses high strength impact wave of the interaction process between high power density (GW/cm^2) lasers and materials to generate residual compressive stress and high-density dislocation, and improves the mechanical properties of metal effectively, especially the metal materials of hardness, residual stress distribution, and resist fatigue life. When the pressure of high strength impact shockwave exceeds the dynamic yield strength of the metal, the shock wave would induce severe plastic deformation in the surface layer of the metals and alloys. Since laser could be easily directed to fatigue-critical areas, LSP is expected to be widely applicable for improving the fatigue properties of alloys. LSP also has the obvious advantages including no contact and heat-affected zone, excellent control ability, and impact.

In the past three decades, LSP has been widely and intensively investigated over 200 scientific papers and reports. Most studies and investigations have been based on experimental approaches, influences of LSP on mechanical properties and, in particular, fatigue lives of metallic materials and alloys. Many researches have been focusing on analytical models and dynamic finite element models (FEM), to simulate the distribution of three-dimensional residual stresses in relation to materials properties, component geometry, laser sources, and LSP parameters in the last decade. LSP is also an effective surface treatment and post-processing method to eliminate tensile residual stress in the surface layer of metallic material and its weldment in order to improve their mechanical properties and tensile performances.

The book is aimed at audiences at different levels to provide a comprehensive discussion of laser shock peening. Both theoretical work and experimental results on LSP are revealed in this book.

Chapter 2 gives a comprehensive review of numerical simulation which could resolve engineering problems and physical problems, and even the nature phenomena by numerical calculation and image displayed method. At present the main method of numerical simulation is the finite element method, the finite difference method, and the finite volume method. Compared with traditional experiment method, numerical simulation has been widely used in many fields, such as mechanical process, large building fire temperature field, hydrogeology, etc. Simulation methods are introduced in this chapter. For instance, the residual stress induced by laser shock processing and the thermal relaxation behaviors of residual stress in Ni-based alloy GH4169 were investigated by means of three-dimensional nonlinear finite element analysis. Fracture analysis software and crack growth model have found application in finite element analysis.

Chapter 3 presents the influences on 00Cr12 alloy's mechanical properties at high temperatures, on metallographic structure evolution and dislocation configuration of 6061-T651 aluminum alloy at elevated temperature, and on ASTM: 410L00Cr12 microstructures and fatigue resistance in the temperature range 25–600 °C.

Chapter 4 focuses on the effect of the compressive residual stresses generated due to laser shock processing (LSP) on the stress intensity factor (SIF) of a through-the-thickness radial crack at the edge of the circular hole, the effect of laser shock peening (LSP) on the fatigue crack initiation and propagation of 7050-T7451 aluminum alloy, and a new kind of statistical data model which described the fatigue cracking growth with limited data and the effects of the reliability and the confidence level to the fracture growth. Many materials have displayed pronounced improvements in fatigue life after LSP. It has shown that LSP treatment improves the materials mechanical properties, fatigue resistance, foreign object damage (FOD), and fatigue life.

Chapter 5 gives a well-rounded presentation of the continuous synthesis of UNCD via laser shock processing (LSP) of graphite particles suspended in water by an Nd: YAG laser system with high power density (10^9 W/cm^2) and short-pulse-width at room temperature and normal pressure, which yielded the ultra-nanocrystalline diamond of size of about 5 nm. X-ray diffraction, high-resolution transmission electron microscopy and laser Raman spectroscopy were used to characterize the nano-crystals. The method studied is helpful in understanding the formation mechanism and enhancing the yield rate of nano-diamond.

Our book entitled Laser Shocking Nano-Crystallization and High-Temperature Modification does not intend to substitute but be a complement to the related books taking advantage of the developments which appeared in my study. It is dedicated to researchers and engineers involved with LSP, its developments, and implementation. The book is also intended for graduate and undergraduate students wanting to be specials in LSP. This book is also conducive to researchers on laser processing, as well as carries great and profound implications for studying laser shocking peening. In this book, the author utilizes the best arguments and the most popular language to explain the theory of laser nano-crystallization.

The research work from which this book arises was carried out at Laser Technology Institute in Jiangsu University supported by the National Natural Science

Foundation of China (Grant No: 51275556, 51479082, 50905080, 51239005). My gratitude goes to many of my colleagues and friends discussing and helping at Laser processing Laboratory, school of mechanical engineering, Jiangsu University.

Zhenjiang, People's Republic of China   Xudong Ren

Contents

1 General Introduction 1
   1.1   Laser Shock Wave and Laser Shock Processing 1
   1.2   Recent Development of Laser Shock Processing 4
   1.3   Practical Applications of LSP 6
   1.4   Scope of This Book 7
   References 8

2 LSP Numerical Simulation 11
   2.1   Introduction 11
   2.2   A Finite Element Analysis of Thermal Relaxation
       of Residual Stress in Laser Shock Processing Ni-based
       Alloy GH4169 12
       2.2.1   Introduction 12
       2.2.2   LSP Simulation 13
       2.2.3   Results and Discussion 16
       2.2.4   Summary 18
   2.3   Comparison of the Simulation and Experimental Fatigue
       Crack Behaviors in the Nanoseconds Laser Shocked
       Aluminum Alloy 19
       2.3.1   Introduction 19
       2.3.2   Simulation Methods 20
       2.3.3   Experimental Methods 22
       2.3.4   Results and Discussions 23
       2.3.5   Summary 29
   References 29

3   Laser Shock Processing at Elevated Temperature.   33
   3.1   Introduction   33
   3.2   Mechanical Properties and Residual Stresses Changing
        on 00Cr12 Alloy by Nanoseconds Laser Shock Processing
        at High Temperatures   34
      3.2.1   Introduction   34
      3.2.2   LSP on Fatigue Properties at Room Temperature   35
      3.2.3   LSP on Fatigue Properties at Elevated
             Temperature   37
      3.2.4   LSP on Residual Stresses at Elevated Temperature . . .   41
      3.2.5   Conclusions   46
   3.3   High-Temperature Mechanical Properties and Surface
        Fatigue Behavior Improving of Steel Alloy Via Laser
        Shock Peening   46
      3.3.1   Introduction   47
      3.3.2   Experiments and Analysis   48
      3.3.3   Results and Discussions.   49
      3.3.4   Conclusions   53
   3.4   Metallographic Structure Evolution and Dislocation
        Polymorphism Transformation of 6061-T651 Aluminum
        Alloy Processed by Laser Shock Peening: Effect
        of Tempering at the Elevated Temperatures   54
      3.4.1   Introduction   55
      3.4.2   Experimental Procedures   56
      3.4.3   Results and Discussions.   58
      3.4.4   Conclusions   74
   References.   74

4   Influence of LSP on Stress Intensity Factor
   of Hole-Edge Crack   79
   4.1   Introduction   79
   4.2   Stress Intensity Factor Changing on the Hole Crack
        Subject to Laser Shock Processing and Influence
        of Compressive Stress   80
      4.2.1   Introduction   80
      4.2.2   SIF Formula.   82
      4.2.3   Experimental Procedures   86
      4.2.4   Analysis Model   86
      4.2.5   Results and Discussion   96
      4.2.6   Conclusions   98

4.3		Investigation of Stress Intensity Factor on 7050-T7451 Aluminum Alloy by High Strain Rate Laser Shock Processing	99
	4.3.1	Introduction	99
	4.3.2	Experiments and Method	100
	4.3.3	Results and Discussions	101
	4.3.4	Conclusions	108
4.4		A Model for Reliability and Confidence Level in Fatigue Statistical Calculation	109
	4.4.1	Introduction	109
	4.4.2	Analysis of Statistics Model	110
	4.4.3	LSP and Fatigue Experiment	112
	4.4.4	Revision of Statistics Model	113
	4.4.5	Conclusions	118
References			118

5   Conversion Model of Graphite ... 123

5.1		Introduction	123
5.2		A Conversion Model of Graphite to Ultra-nano-crystalline Diamond via Laser Processing at Ambient Temperature and Normal Pressure	124
	5.2.1	Experiment and Method	124
	5.2.2	Results	125
	5.2.3	Discussion	127
	5.2.4	Conclusions	130
References			130

About the Author

Dr. Xudong Ren is an associate professor at School of Mechanical Engineering, Jiangsu University, People's Republic of China. He received his Ph.D. degree at Jiangsu University (P.R.China) in 2009. His research includes laser shock processing/peening, laser shocking nano-crystallization, and high-temperature modification and laser compound processing.
Email: renxudong@vip.163.com; renxd@mail.ujs.edu.cn

Chapter 1
General Introduction

Abstract Laser shock processing (LSP) is a new surface modification technology which generates high-strength impact wave and then introduces compressive residual stress of several hundred MPa by exposing metallic samples to high-power density (GW/cm^2), short-pulse (ns level) laser beam. This chapter presents recent development of LSP, the LSP process, practical applications of LSP, and the scope of this book.

1.1 Laser Shock Wave and Laser Shock Processing

Laser surface treatment is a flexible way of effectively protecting component and tool surfaces from wear and corrosion. Various surface treatment techniques can prolong tool life and improve component, tool, or die performance. During LSP, the action on the surface and subsurface of the materials is achieved through the mechanical effect arisen from the laser shock wave, as schematically shown in Fig. 1.1.

Laser surface treatment is a novel and potential subject of considerable interest present because it seems to offer the chance to save strategic materials or to allow improved components with idealized surfaces and bulk properties [2]. Laser surface treatment shows many advantages compared to conventional techniques, e.g., a high flexibility with respect to the processed geometries or the possibility for a simple integration into existing production lines. In particular, when only a small part of the surface of a work piece shall be treated, the laser process should be preferred.

Laser surface treatment can cause the changes in surface properties of alloys and metallic materials by generating temperature gradients, phase changes, or mechanical influences [3]. The most important processes are shown as following:

Laser shock processing shock hardening occurs when laser pulses with duration in the ns range are applied. In this case, shock waves are induced, which cause a

© Springer-Verlag Berlin Heidelberg 2015
X. Ren, *Laser Shocking Nano-Crystallization and High-Temperature Modification Technology*, DOI 10.1007/978-3-662-46444-1_1

Fig. 1.1 Laser shock
processing principle and
schematic. Reprint from Ref.
[1], Copyright 2013, with
permission from Elsevier

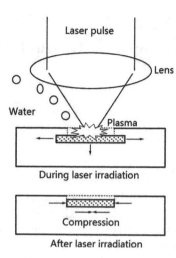

kind of mechanical deformation connected to an increase of the hardness. The
involved mechanisms are therefore similar to those occurring during cold working.

The LSP process utilizes high-energy laser pulses (several GW/cm^2) fired at the
surface of a metal covered by two layers, namely an absorbing layer and a water-
confining layer. When a laser pulse with sufficient intensity passes through the
transparent confining layer and hits the surface of the material, the absorbent
material vaporizes and forms plasma. The plasma continues to strongly absorb the
laser energy until the end of the energy deposition. The rapidly expanding plasma is
trapped between the sample and the transparent confining layer, creating a high
surface pressure, which propagates into the material as a shockwave. When the
pressure of the shockwave exceeds the dynamic yield strength of the metal, it
produces plastic deformation in the near surface of the metal.

The LSP technology has the following advantages: (1) less surface roughening
compared to the conventional shot peening; (2) no embedded particles; (3)
strengthening right into corners where shot could not reach; (4) no material to
recycle, collect, grade, and clean, as there is with shot peening; and (5) flexible laser
pulse beam which can be adjusted and controlled in real time.

The high-intensity laser can induce shock wave to change the mechanical
properties and microstructure of 7075 aluminum alloy, which was discovered by
Fairand B.P. in the Unite Battelle Memorial Institute in the early 1970s for the first
time. The yield strength of 7075 aluminum alloy increased obviously, and countries
launched on the study of LSP technology [4]. Now, the world's most advanced
research institutions in the field of LSP research and application contents are
General Electric Aircraft Engines (GEAE), LSP Technologies, Inc. (LSPT), Lawrence Livermore National Laboratory (LLNL), and Metal Improvement Company
(MIC) [5].

Gen I, the first generation of GEAE LSP equipment from 1992 to 1994: GEAE
used laser equipment produced by Battelle Memorial Institute to improve

high-cycle fatigue (HCF) properties of engine blades and foreign body damage (FOD) resistance by LSP, and verified its effectiveness. The laser machine equipped eight independent laser heads, and typical parameters are as follows: pulse energy of 100 J, pulse width of 30 ns, and repetition frequency of 0.1 Hz. GEAE got the laser machine by lease and named it as "Gen I." GEAE utilized this equipment to do the preliminary study on the application of LSP. Then, the Gen I laser system and multifreedom manipulators, which can be accurately controlled by computer, were applied to industrial production in 1997.

Gen II, the second generation of GEAE LSP equipment: GEAE designed and built the second generation of LSP system-Gen II through LSPT, as shown in Fig. 1.2. The reasons for the design of this system are the low impact frequency of Gen I limiting productivity. Typical output parameters of Gen II which are equipped with 12 independent laser heads are as follows: 100 J, 30 ns, and 0.25 Hz. Three-year application proved that the Gen II has a good stability, and the overhaul cycle of the system increased from 3 to 12 months compared with Gen I. But the maintenance cost is still high due to the complex structure, and the cost of each system is more than 100,000 dollars per year. The cost mainly concentrated on the replacement of large size of lens (Φ15 mm) and material glass rods (Φ4.5 mm).

Gen III, the third generation of GEAE LSP equipment: This laser machine is different from Gen I and Gen II, and the typical character is the lath structure rather than a bar structure. The usage of lath structure gives the Gen III output enough energy with one laser head, one oscillator, and two amplifiers. However, the reliability of the Gen II is quite high, which is enough to be applied to production, so the Gen III is not used on the practical production but on further research of LSP.

Gen IV, the fourth generation GEAE LSP equipment: The limitations of the first three generations of LSP system are the complicated technology and the expensive price. Compared with equipment which is applied to the laser drilling, the price is too expensive. A strong maintenance team is needed because of technical

Fig. 1.2 Gen II: the second generation of GEAE LSP equipment

Fig. 1.3 Gen IV: the fourth
generation of GE LSP
equipment [6]

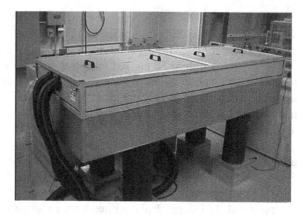

complexity, so GE established LSP production lines both in Cincinnati and Ohio to facilitate routine maintenance of LSP system.

The strict using requirements of LSP system prompted GE to begin the research on low-energy laser with a simple structure and low cost to achieve the same strengthening effect. In 2000, the GEAE engineers reassembled a Gen I laser and made the laser output several joules in a pulse. After debugging, the surface of Ti alloy generated a significant dent after laser shock processing by this equipment, which means a certain level of residual compressive stress was produced on the surface of the material. It is proved that the low-energy laser machine (minimum to 5 J) can produce 1.5-mm-deep residual compressive stress layer on the specimen surface, which is in the same level of Gen I and Gen II. This kind of low-energy laser is called Gen IV, as shown in Fig. 1.3.

Gen IV is applied to the treatment of the fan blade F101-GE-102 engines, so as to assess the deformation degree of the blade shocked by laser and the improvement of HCF life. Experiments showed that in these two indicators, Gen IV can achieve the same level as the Gen II. The results of the study prompted the GEAE to develop the commercial use of low-energy and high-repetition-frequency laser system (10 J, 30 ns, 10 Hz) which can reduce the cost and significantly improve the productivity.

1.2 Recent Development of Laser Shock Processing

In the early 1960s, some researchers have found that using pulsed laser on material surface can produce high-strength shock waves in the solid [7–9], but there is no research on the stress shock used for material modification. Until 1972, in Unite States Battlell's Columbus Lab, Fairand and other researchers changed the microstructure and mechanical properties of 7075 aluminum alloy with high-power pulsed laser-induced waves for the first time. The result shows that the yield strength $\delta_{0.2}$ of 7075 aluminum alloy increased by 30 % [10] after laser shock processing. The laser has good controllability, repeatability, and many other features, so the shock waves produced by pulse laser become a new tool of surface modification of materials, which has opened up the prolog of laser shock processing for materials applied research.

High-power Nd:YAG laser device is commonly used in laser shock processing [11]. At present, large high-power laser devices generally use neodymium glass laser at home and abroad. The pulse width of neodymium glass laser is 10^{-9} s or even smaller, so the laser power density is generally greater than 10^9 GW/cm^2, sometimes up to 10^{13} W/cm^2. When such a high-power density and short-pulse laser acts on the material surface, the thin layer of material surface gasifies rapidly. Momentum pulse occurs during surface atomic escaping and high-pressure shock wave as high as several GPa, even up to 15 TPa [12, 13], is produced. As the shocking time is significantly short, the work piece is hardly affected by the heat and the surface temperature is only around 150 °C, the laser shock processing mainly uses the high-pressure mechanical effect but thermal effect. At this point, laser shock processing can be classified as cold working process [14, 15].

Using the mechanical effect of laser shock wave on the material surface is called laser shock technology. Taking advantage of laser shock wave on the material surface for modification to improve materials' fatigue, stress corrosion and wear resistance attracts researchers' attention both at home and abroad. This technology is currently recognized as the "Laser Shock Processing" or "Laser Shock Peening." Extremely high shock pressure (up to 10 TPa), which is the highest shock pressure obtained in laboratory, can be produced by laser shock processing. Laser shock processing is mainly applied to metal materials to enhance mechanical properties.

In 1995, Jeff Dulaney established the world's first laser shock processing technology company (LSP Technologies Inc.), set up the laser processing Web site, and registered the LSP trademark [16]. The company provides the whole industry with laser shock processing technology and processing services relying on its advanced laser shock processing equipment and systems.

In June 2002, the American Lawrence Livermore National Laboratory (the most authoritative and highest level laboratory in the research on high-power Nd glass laser device), cooperating with United States Metal Modification Company (MIC), introduced the laser shock forming and laser shock processing system, which had been performed on the slice of turbine engine, and applied for a patent for this invention [17, 18]. The main performance indicators of this system are the following: Work material is neodymium glass in the shape of plate strip; laser pulse energy 50 J; laser pulse width (FWHM) 10–30 ns; and practical application working repetition rate 0.5–1 Hz. However, this system is too large and expensive.

Laser shock processing can produce high compressive residual stress on the surface of metallic materials, which would significantly improve materials' mechanical properties. However, some investigations reveal that residual stress would release at elevated temperatures. Study on the relaxation mechanism of superalloys that are serving in the environment with high temperature and bearing fluctuation loading is significantly important for improving service life. Besides, during laser shock processing, residual stress in the surface changes and stress intensity factors also show different characteristics at various stages of the shock. Several studies have been done on the effects of residual stress on stress intensity factors. But for laser shock process, stress intensity factors changing under impact surface and in the process of fatigue crack initiation and propagation have not been elaborated, which also prevented the further development of the study on surface cracks induced by laser. Laser shock processing can produce compressive residual stress on material surface and make changes in the microstructure of the surface, even nano-crystals. Nano-crystallization induced by laser shock wave differs from other mechanical methods, which highlights the advantage of unchanged surface roughness and better thermal stability. Laser nano-crystallization is the new direction and challenge of laser shock processing, and its practical application still needs further study.

1.3 Practical Applications of LSP

LSP attracts strong interest in the field of commercialization application. Since two important patents were first issued in 1974 and 1983, LSP has been gradually used in industry. From 1996, the General Electric Company alone applied for a large number US patents based on laser shock processing.

In the aerospace industry, laser shock processing is an effective method to improve the mechanical properties and fatigue lives of aerospace key products, such as turbine blades [19] and rotor components [20, 21], disks, gear shafts [22], and bearing components [23]. Laser shock processing could also be applied to strengthen fastener holes in cover parts. GEAE in the USA treated the leading edges of turbine fan blades [19] in F101-GE-102 turbine for the Rockwell B-1B bomber by laser shock processing in 1997, which enhanced fan blade durability and resistance to foreign object damage (FOD) without harming the surface finish [18]. Protection of turbine engine components against FOD is a key priority of the US Air Force. In addition, it was reported that laser peening would be applied to treat engines used in the Lockheed Martin F-16C/D [24].

LSP Technologies Incorporation (Dublin, OH) also recently commissioned the ManTech Laser Shock Peening Manufacturing Cell (LSPMC). From 2004, LSP Technologies laser peened the airfoils on the Pratt and Whitney F119, 4th stage IBR, that was flown on the F/A-22 Raptor aircraft. Implementation of laser shock processing increased the notched fatigue strength of IBR airfoils above the 55ksi fatigue strength design criteria [25]. The application of laser shock processing to the F119 IBR has reduced maintenance costs and eliminated the need for a costly redesign, estimated to be greater than $10 M.

Laser shock processing can obviously improve the fatigue strength of the damaged blades which is equal to or better than that of undamaged blades [26]. Laser shock processing has been shown to provide more than 2.5 times lifetime enhancement against fatigue failure for coupons replicating T-45 arrestment hook shanks. Employment of laser shock processing on aircraft would reduce maintenance costs and add to aircraft availability [27]. Since 2002, Metal Improvement Company (MIC) has processed over 35,000 wide-chord fan blades for commercial aircraft, and laser shock processing can extend lifetime of new and used Boeing 777 blades by more than 20 times.

Laser shock processing without protective coating (LSPWC) was also developed and the practical effects (SCC and fatigue prevention). This technology has been applied to Japanese nuclear power reactors (BWRs and PWRs) as preventive maintenance against SCC since 1999 [28].

1.4 Scope of This Book

Mechanical surface treatment, such as shot peening and deep rolling, are known to induce several beneficial effects into metallic surfaces, which serve to enhance fatigue properties. Compared with the traditional process such as shot peening, deep rolling, and cold extrusion, LSP is a new surface modification technology to improve the mechanical properties of metal, especially the metallic material of hardness, residual stress and fatigue life. The strengthening mechanism of LSP on the components at elevated temperature would have a difference as compared with room temperature. The aim of Chap. 2 is to study the influences of elevated temperature on surface topography, micro-hardness, and microstructure of different metallic materials by LSP. Furthermore, the strengthening mechanisms of LSP on micro-hardness and dislocation configuration evolution at elevated temperature are also revealed.

In most cases, the residual stress relaxation of surface-treated alloys is studied through experimental trials which are expensive, time-consuming, and unreliable resulting from factors arising from setting up the LSP process and residual stress measurements [29]. Recently, the finite element (FE) method has been used to study LSP-induced temperature and residual stress distribution in alloys. FE models have been established in Chap. 2 to study the residual stress distribution induced by LSP and the stress relaxation behaviors under thermal loading condition.

Chapter 3 introduces high-temperature modification of metallic materials after LSP. Temperature has a significant effect on surface topography and mechanical properties of LSP specimens. In this chapter, three different kinds of alloy were studied to investigate the relationship between mechanical properties and temperature. Besides, metallographic structure of LSP materials varies at elevated temperatures, which is also involved in Chap. 3.

Chapter 4 examines the effect of laser shock processing on stress intensity factor, analyzing stress intensity factor changing on the hole crack subjected to laser shock processing and influence of compressive stress. A model for reliability and confidence level in fatigue statistical calculation is given in this chapter. The effects of stress intensity factor on 7050-T7451 Al alloy by high-strain rate laser shock processing were investigated as well.

Chapter 5 puts forward the transformation model and growth restriction mechanism of high power density with short-pulsed laser shocking of graphite particles in liquid. A conversion model of graphite to ultra-nano-crystalline diamond has been established in this chapter, and the detections of XRD, HRTEM, and Raman spectroscopy were implemented to identify the transformation of graphite particles into UNCD as well.

References

1. Ren XD et al (2013) The effects of residual stress on fatigue behavior and crack propagation from laser shock processing-worked hole. Mater Des 44(2013):149–154
2. Sahin AZ et al (2014) Laser surface treatment and efficiency analysis. Compr Mater Process 9:307–316
3. Yilbas BS (2014) Laser treatment of boron carbide surfaces: metallurgical and morphological examinations. Compr Mater Process 9:1–3
4. Fairand BP, Clauer AH (1979) Laser generation of high-amplitude stress waves in materials. Appl Phys 50(3):1497–1502
5. Chu JP et al (1995) Effects of laser-shock processing on the microstructure and surface mechanical properties of Hadfield manganese steel. Metall Mater Trans A Phys Metall Mater Sci 26A(6):1507–1517
6. Davis BM et al (2014) Performance of Gen IV LSP for thick section airfoil damage tolerance. AIAA 2062:1–10
7. Fabbro R et al (1986) Experimental study of metallurgical evolutions in metallic alloys induced by laser generated waves. In: Society for optical engineering, vol 668. Quebec City Que, Can, SPIE, Bellingham, WA, USA, pp D320–324, 3–6 Jun 1986
8. William F, Bates Jr (1997) Laser shock processing aluminum alloy application of laser in material processing. American Society for Metals (ASM), Washington DC, 18–20 Apr 1979
9. Ford SC et al (1980) Investigation of laser shock processing-executive summary technical report AFWAL-TR-80-3001, Voll, Aug 1980
10. Fairand BP et al (1972) Laser shock induced microstuctural and mechanical property changes in 7075 Aluminum. J Appl Phys 43(9):3893–3895
11. Zhang Y (2005) Laser shock wave loading metal materials under high strain rate forming reinforcement technology and device technical summary report. Jiang Su University, Dec 2005
12. Masse JE, Barreau G (1995) Laser generation of stress waves in metal. Surf Coat Technol 70:231–234
13. Romain JE et al (1986) Laser shock experiments at pressure above 100 Mbar. PhysicaB: physica of condensed Matter&C: Atomic. Molecular and plasma physica. Optics 139–140:595–598
14. Jianzhong Z, Yongkang Z (2000) Application of laser processing technology in automobile body manufacturing. Electr Process Mould 4:32–36
15. Vaccari JA (1992) Laser shocking extends fatigue life. Am Machinist 7:62–64
16. http://www.lsptechologies.com/applicant.asp. Accessed 17 Mar 2014
17. Yeaton RL (1998) Method and apparatus for laser shock peening. US Patent 5744781, Apr 1998
18. Clauer AH, Firand BP (1998) Laser peening process and apparatus. US Patent 5471599, Apr 1998
19. Zhang Y (1997) The effect of laser shock processing intuitive criterion and control method research. Laser Chin A24(5):467–471

20. Lan C, Zhang Y (1996) Laser shock resistance to metal fatigue fracture of laser shock parameters optimization test research. Laser Chin A 23(12):117–120

21. Jianzhong Zhou et al (2001) Laser impact effect on the properties of ductile iron surface. Appl Laser 2:91–95

22. Ferrigno SJ (2001) US Patent 6,200,689. General Electric Company, Cincinnati, OH, 13 Mar 2001

23. Casarcia DA et al (1996) US Patent 5,584,586. General Electric Company, Cincinnati, OH, 17 Dec 1996

24. Brown AS (1998) A shocking way to strengthen metal. In: Aerospace America, pp 21–23

25. Sokol DW et al (2004) Applications of laser peening to titanium alloys. In: The ASME/JSME 2004 pressure vessels and piping division conference, San Diego CA, 25–29 July 2004

26. http://www.lsptechologies.com/FatRevChart.pdf. Accessed 17 Mar 2014

27. Rankin J et al (2010) Effect of laser peening on fatigue life in an arrestment hook shank application for naval aircraft. In: The 2nd international laser peening conference, San Francisco CA, 19–22 Apr 2010

28. Sano YJ et al (2010) Applications of laser peening without protective coating to enhance structural integrity of metallic components. In: The 2nd international laser peening conference, San Francisco CA, 19–22 Apr 2010

29. Achintha M et al (2013) Eigenstrain modelling of residual stress generated by arrays of laser shock peening shots and determination of the complete stress field using limited strain measurements. Surf Coat Technol 216:68–77

Chapter 2
LSP Numerical Simulation

Abstract This chapter gives a comprehensive review on numerical simulation which could resolve engineering problems and physical problems even the nature phenomena by numerical calculation and image displayed method. At present, the main method of numerical simulation is the finite element (FE) method, the finite difference method, and the finite volume method. Compared with traditional experiment method, numerical simulation has been widely used in many fields, such as mechanical process, large building fire temperature field, and hydrogeology. Simulation methods are introduced in this chapter. For instance, the residual stress induced by laser shock processing (LSP) and the thermal relaxation behaviors of residual stress in Ni-based alloy GH4169 were investigated by means of three-dimensional nonlinear FE analysis. Fracture analysis software and crack growth model have found the application in FE analysis.

2.1 Introduction

At present, the main method of numerical simulation is the finite element (FE) method, the finite difference method, and the finite volume method. The FE method divides the continuous domain decomposition field into a finite number of units, consisting of discrete model, and then finds its approximate numerical solution. The finite difference method establishes a set of differential equations taken the place by difference to cover the space and time with the grid for approximate numerical solution. The finite volume method transforms differential equations into integral form in a physical space and scatters it. After theoretical analysis and scientific experiments, numerical simulation has become one of the main means of science and technology development.

Numerical simulation is also called computational simulation which could resolve engineering problems and physical problems, even the nature phenomena, by numerical calculation and image displayed method. Compared with traditional experiment method, numerical simulation has been widely used in many fields,

© Springer-Verlag Berlin Heidelberg 2015

X. Ren, *Laser Shocking Nano-Crystallization and High-Temperature Modification Technology*, DOI 10.1007/978-3-662-46444-1_2

such as mechanical process, large building fire temperature field, and hydrogeology. The experimental results can be predicted by simulating the process of actual reaction with appropriated numerical simulation software, greatly saving time and manpower consumption. LSP is that the laser of high power and short pulse radiates metal surface in order that the coat of it gasifies to the high-temperature and high-pressure plasma. Compared with traditional surface strengthening, LSP possesses the advantages of no mold, easily controlling, highly flexible processing, and good surface properties after strengthening and no pollution. Therefore, it is impossible to study LSP process only relying on experiment. Numerical simulation as a kind of supplementary tool will accelerate the development and application of LSP in mechanical engineering.

2.2 A Finite Element Analysis of Thermal Relaxation of Residual Stress in Laser Shock Processing Ni-based Alloy GH4169

Abstract The residual stress induced by LSP and the thermal relaxation behaviors of residual stress in Ni-based alloy GH4169 were investigated by means of three-dimensional nonlinear FE analysis. To study the effect of different given exposure time and different temperatures on residual stress in LSP Ni-based alloy GH4169, Johnson–Cook material model was used in order to account for the nonlinear constitutive behavior. The influence of heating temperature and exposure time on the stress thermal relaxation was studied. It was concluded that stress relaxation mainly occurred during the initial period of exposure, and the degree of relaxation increased as the temperature risen. The results would provide a theoretical basis for controlling the LSP and guiding subsequent experiments.

2.2.1 Introduction

Ni-based alloy, such as GH4169, is widely used in high serving temperature fields such as space navigation, nuclear energy, and petroleum industry due to excellent fatigue resistance, radiation resistance, corrosion resistance, good machinability, and welding performance. Various mechanical surface treatment technologies, such as shot peening [1], low plastic polishing [2], and LSP [3, 4], have been used to restrain the propensity of fatigue initiation or growth by inducing compressive stress on the surface and fellow-surface regions to improve the fatigue performance of metallic materials. Compared with other surface treatment technologies, LSP has unique advantages that it could be performed contactless with component and without heat-affected zone.

The temperature and exposure time are the primary parameters for the thermal relaxation of residual stress. Masmoudi et al. [5] have studied the thermal relaxation of residual stress in shot peened Ni-based alloy IN100 as exposed to different temperatures (500–750 °C). During the initial period of exposure, the surface residual stress decreased rapidly. Cao et al. [6] also observed similar phenomena. Khadhraoui et al. [7] performed an experiment to investigate thermal stress relaxation in IN718 with different exposure times at 600 and 650 °C. Prevey et al. [1] studied the thermal stress relaxation of the compressive layer produced by LSP. Cai et al. [8] studied the residual compressive stress field of IN718 induced by shot peening and the relaxation behavior during aging, and the relaxation process was described by the Zener–Wert–Avrami function [9, 10].

In most cases [11], the residual stress relaxation of surface treated alloys is studied through experimental trials, which are expensive, time-consuming, and unreliable resulting from the factors arising from setting up the LSP process and residual stress measurements. Recently, the FE method has been used to study LSP-induced residual stress in titanium alloys [9], and the simulation results are well consistent with the experimental results. However, there is little FE analysis on thermal relaxation of residual stress having been performed. The purpose of this chapter is to study the residual stress distribution in GH4169 induced by LSP and the stress relaxation behaviors under thermal loading condition.

2.2.2 LSP Simulation

2.2.2.1 Thermal Relaxation

High-pressure shock waves are induced after laser having an impact on the absorption layer and constraint layer attached to the surface of components, which propagates into the material and causes plastic deformation resulting to residual stress field. In order to accurately simulate the LSP and the subsequent thermal relaxation process, the material constitutive model must accord with the properties of the material and the deformation of the FE model. During the LSP, material experiences extremely high strain rate reaching the order of $10^6 \, \text{s}^{-1}$. In addition, the flow stress would reduce at elevated temperature, and the temperature effect is thought to be the primary reason causing the thermal relaxation of residual stress. A lot of material models have been developed to explain the effect of strain hardening, strain rate hardening, and thermal softening. These models include Johnson–Cook (JC) [9, 10, 12], Khan–Huang–Liang (KHL) [9], Cowper–Symonds [13], Zerilli–Armstrong (ZA) [14], and optimization of ZA. In this study, JC model is employed for that it is widely used for impact applications and is available in material model library in many commercial FEM codes, such as ANSYS/LSDYNA which is used in this study. Coupled structural/thermal implicit analysis is performed for thermal stress relaxation [9, 10]. JC model is given as

$$\sigma = [A + B(\varepsilon^p)^n][1 + C\ln(\varepsilon^{-p}/\varepsilon_0)][1 - T^{*m}] \tag{2.1}$$

where σ is the flow stress; ε^p is the effective plastic strain; ε^{-p} is the effective plastic strain rate, and ε_0 is reference strain rate; A, B, C, n, and m are experimentally determined constants. $T^* = (T - T_r)/(T_m - T_r)$, where T is the temperature in Kelvin, and T_r and T_m are room temperature and melting temperature, respectively.

2.2.2.2 Experimental Materials and Parameters

For the purpose of saving the memory of computer and shortening the computation time, only a quarter of models are created. The energy of laser is 8 J, the diameter of spot is 3 mm, and the pulse width is 10 ns. The FE model with 8-noded solid element is shown in Fig. 2.1, which has 178,596 nodes and 169,000 elements. As Ni-based alloy GH4169 and IN718 super alloy are similar materials, the original JC material model parameters for GH4169 ($A = 900$ MPa, $B = 1200$ MPa, $C = 0.0092$, $n = 0.6$, $m = 1.27$, $\varepsilon_0 = 1s^{-1}$) are taken from Zhong et al. [10]. After optimization, the JC material model with constants $A = 860$ MPa, $B = 1100$ MPa, $C = 0.0082$, $n = 0.5$, $m = 1.05$, and $\varepsilon_0 = 1s^{-1}$ is used for this study.

The pressure waveform of laser shock wave is similar to the laser wave profile tested by oscilloscope, according to Fabbro et al.'s findings [14, 16]. The studies have shown that the response time of laser inducing shock wave is about 3 times of the laser pulse width, or even longer. In this study, as the laser pulse width is 10 ns, the response time is set as 70 ns. The temporal pressure pulse profile is shown in Fig. 2.2b.

The residual stress from FE simulation is plotted in Fig. 2.3. Figure 2.3 shows the distributions of the residual stress with laser energy 6, 8, 12 and 15 J,

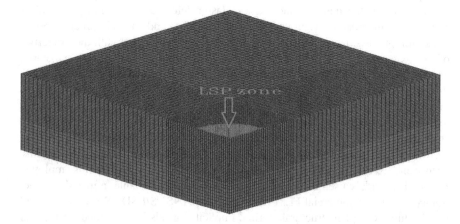

Fig. 2.1 FE meshes of 10, 10, and 5 mm for LSP and thermal relaxation analysis. Reprint from Ref. [15], Copyright 2014, with permission from Elsevier

Fig. 2.2 a Amplitude curve of shock pressure; **b** Temporal pressure pulse profile. Reprint from Ref. [15], Copyright 2014, with permission from Elsevier

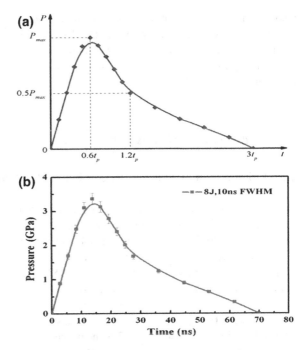

Fig. 2.3 Residual stress induced by LSP with 6, 8, 12 and 15 J. **a** Surface residual stress fields with single impacts; **b** Residual stress fields along with the depth with single impacts. Reprint from Ref. [15], Copyright 2014, with permission from Elsevier

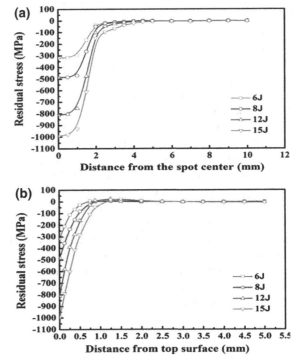

respectively. It could be observed that the maximum compressive stress increases as the laser energy rises. And the same trend is shown in the depth of compression layer. A higher magnitude of stress is obtained for greater laser energy. The JC model has good agreement with experimental results, indicating the well consistency of the model [4].

2.2.3 Results and Discussion

2.2.3.1 Simulation of Thermal Relaxation of Residual Stress

The FE model for simulating the thermal relaxation of residual stress induced by LSP is the same as the one used in the LSP analysis. The mechanical loading is removed, and the thermal loading is applied at the top, bottom, and the both sides of the surface to simulate the thermal relaxation process. This process is simulated through a coupled thermal–structure analysis using ANSYS/LS-DYNA, with implicit algorithm for both thermal and structure analysis. The purely thermal loading condition is simulated by using a convection boundary condition $q = h(T - T_\infty)$, where q is heat flux across the boundary, h is heat transfer coefficient [100 W/(m^2 °C)], T_∞ is the imposed heating temperature, and T is the initial temperature of specimen. The initial temperature of the Ni-based alloy GH4169 specimen before applying thermal loading is set as the room temperature. The heat capacity of GH4169 is 481.4 J/(kg °C), and the thermal conductivity is 13.4 W/(m °C).

The residual stress induced by LSP with laser energy of 8 J and its relaxation as a function of exposure time at 500, 600 and 700 °C are shown in Fig. 2.4. The result reveals that only a little relaxation occurred at temperatures. It is expected that a greater degree of thermal relaxation would take place when the temperature becomes high enough to reduce the yield stress of GH4169. Another mechanism of thermal relaxation arises when the residual stress reaches sufficiently high values as to exceed the yield stress at the temperature of interest [10].

Figure 2.4a, b also shows that the maximal relaxation occurred at the zone of maximum residual stress. Stress relaxation mainly occurs in the initial exposure period, then the residual stress tends to stabilize at a longer exposure time, and eventually the relaxation essentially stops. The results are consistent with those studies [6, 17–19], which indicates a large portion of total relaxation of residual stress occurring in the initial period of exposure (normally between 3 min and 1 h) and following by a stabilization of stress. The effect of temperature on the thermal relaxation behavior is also showing in Fig. 2.4. The amplitude of the thermal relaxation increases with an increase in temperature. In addition, the effect of temperature is more significant at longer exposure time. Similar phenomenon could also be found in Zhong et al. [9, 10].

Fig. 2.4 The effect of temperature on the thermal relaxation behavior; **a** Thermal stress relaxation in GH4169 at different temperatures with a constant exposure time; **b** Thermal stress relaxation in GH4169 at different exposure time with a constant temperature. Reprint from Ref. [15], Copyright 2014, with permission from Elsevier

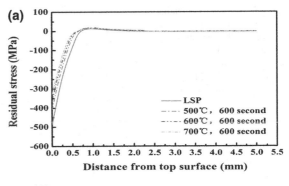

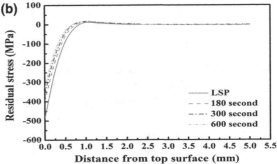

2.2.3.2 Analytical Model for Thermal Relaxation in GH4169

The residual stress is simulated with different exposure times at a constant high temperature and different high temperatures at a constant exposure time. The main work of this study focuses on the effect of temperature and exposure time on thermal stress relaxation.

A stress relaxation model by employing the Zener–Wert–Avrami function [9, 10, 20–22] was proposed to provide an analytical description of the thermal relaxation of residual stress,

$$\frac{\sigma^{RS}}{\sigma_0^{RS}} = \exp[-(At_a)^m] \tag{2.2}$$

where, σ_0^{RS} is the initial value of residual stress prior to apply thermal loading; σ^{RS} is the residual stress at a given time t_a at temperature T_a; m is a numerical parameter dependent on the dominant relaxation mechanism; and A is a function dependent on the material and temperature according to,

$$A = B \exp\left[-\frac{\Delta H}{KT_a}\right] \tag{2.3}$$

Fig. 2.5 Influence of heating
temperature and exposure
time on the thermal stress
relaxation of GH4169.
Reprint from Ref. [15],
Copyright 2014, with
permission from Elsevier

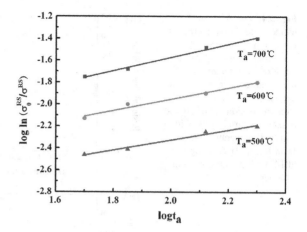

where B is a constant, K is the Boltzmann constant, and ΔH is the activation
enthalpy for the relaxation process in the t_a–T_a range under consideration.

From Eq. (2.2), a plot of $\log \ln \left(\sigma_0^{RS} / \sigma^{RS} \right)$ versus $\log t_a$ at a given aging temperature T_a gives a straight line with slope m and intercept $m \log A$ shown in
Fig. 2.5. The m value resulting from data fit at different temperatures gives a mean
value of $m = 0.68$. The intercept value of these straight lines gives the value of A as
a function of T_a, so that it is possible to estimate the activation enthalpy (ΔH) of the
relaxation process for the LSP GH4169 referring to Eq. (2.3). The calculated value
of ΔH is 3.02 eV, which is quite close to the value of 2.88 eV reported by Zhong
et al. [10] for pure Ni using ion beam sputtering technique.

Dynamic recovery is the main relaxation mechanism during the stress relaxation
between 400 and 500 °C in LSP 6061-T651 aluminum alloy [22]. Essentially,
dynamic recovery is the result of the dislocation slip, the dislocation climb, and the
reducing of dislocation density [22]. We conjecture that it is the creep-like mechanism involving rearrangement and annihilation of dislocations by climbing that
causes the thermal relaxation. After the initial stress relaxation, subsequent stress
relaxation becomes considerably lower, as the stress has dropped below the creep
strength of the alloy at temperatures, and therefore, the driving force is reduced.
However, further metallographic structure studies are needed to verify this, which is
beyond the scope and purpose of this paper.

2.2.4 Summary

The residual stress in the Ni-based alloy GH4169 induced by LSP and the asso-
ciated thermal relaxation behaviors were analyzed based on a three-dimensional
nonlinear FE model.

At higher temperatures, the maximum thermal relaxation occurs at the depth of the maximum compressive residual stress. A significant decrease of the residual stress is observed at the initial exposure period followed by a stabilization of the stress relaxation at the next long exposure times.

The degree of thermal relaxation increases as the temperature rises, and the effect of temperature is more significant at longer exposure times.

Further work should be carried out to study the material model and methodology in overlap case using the single shot. Furthermore, to further study the effect of plastic deformation, the thermal relaxation behavior of doubly shocked and triply shocked specimens at a constant temperature will be simulated.

2.3 Comparison of the Simulation and Experimental Fatigue Crack Behaviors in the Nanoseconds Laser Shocked Aluminum Alloy

Abstract This chapter was performed to compare the simulation and experimental results of the fatigue crack growth rates and behaviors of the 7050-T7451 aluminum alloy by nanoseconds LSP. Forman–Newman–deKoning (FNK) model embedded in the Franc2D/L software was utilized to predict fatigue crack growth rate, which was conducted to weigh the stress intensity factor (SIF) changing on the surface cracks. LSP induced high compressive residual stresses that served to enhance fatigue properties by improving the resistance against fatigue crack initiation and propagation. The circulating times of crack growth obtained from the simulation and experimental values indicated a slower fatigue crack growth rate after LSP. The relationships between the elastic–plastic material crack growth rates and the SIF changing after LSP are resolved.

2.3.1 Introduction

Fracture mechanics typically offers a reliable foundation for the description of the fatigue growth of cracks. It is well known that residual stress plays a crucial role in fatigue crack growth behavior [23–26]. LSP is a competitive technology as a method of imparting compressive residual stresses into the metal surface to improve fatigue and corrosion properties [27, 28].

The research on stress shock wave on cracks has made great progress in aspects of theoretical and numerical simulations. A numerical model for predicting the depth of plastic deformation and resulting residual compressive stress at the surface has been completed by ArifAbulFazal [29]. The plastic behavior near the tip of stationary crack in engineering materials has been intensively studied by using classical plasticity theory based on the von Mises yield criterion [30] and the associative flow rule [31]. Ray and Patanker [32] derived the crack closure models on the crack opening stress by FE computations. Then, the theoretical analysis work which confirmed the effect of the compressive residual stresses generated by LSP on the SIF has been put forward [33, 34], showing the influence of compressive stress on the 3D non-through hole-edge crack's SIF after LSP.

However, the relationships between the fatigue crack growth rates and the surface SIF changing after LSP have never been reported in literature to the authors' best knowledge. With respect to the continuity information between this study and the available literature, we aim to characterize the effects of residual stress on fatigue crack propagation under variable laser shock amplitude loading. Moreover, we will also discuss simulation results on the dynamic response of cracks under the actions of various shock loads, and we use the crack growth function of Franc2D/L to present the curve of the surface SIF changing under residual stress loads. Further studies on the issue of optimal fracture with the action of stress pulses will be summarized in our next study.

2.3.2 Simulation Methods

2.3.2.1 Fracture Analysis Software

Franc2D/L is the professional fracture analysis software based on FE analysis and developed by Cornell Fracture Group. Franc2D/L initialized cracks and predetermined the initiation spots of cracks. Then the software judged the cracks growing direction so as to perform numerical simulation of crack growth and obtain results automatically. Figure 2.6 shows the sample of the singularized element growing along with the crack tip. The crack tip has been modeled with a rosette of eight quarter-point collapsed elements. The propagation of the crack is divided into many steps, and in every step, the crack is let to extend to a certain amount. The first propagation step is set to 0.5 mm, while the following crack propagation steps are 1 mm each. After each step, Franc2D/L remeshes the region in front of the crack tip and calculates the SIF and the kink angle. The crack propagation direction is determined according to the maximum hoop stress criterion (MHSC) [35, 36], and K_I and K_{II} are determined at every step.

2.3.2.2 Crack Growth Model

The FNK model was used to determine the crack growth rate, which could account for retardation near threshold, acceleration near fast fracture, and the crack closure. This method is given as [38],

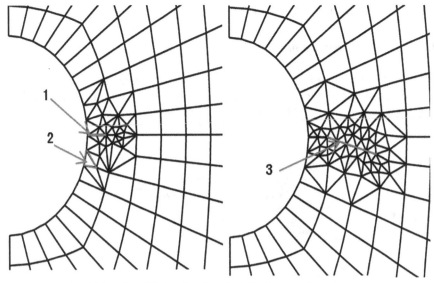

1- Initial crack; 2- T6 triangular elements; 3- Crack propagation path

Before Crack Expanding   **After Crack Expanding**

Fig. 2.6 Sample of the singularized element growing along with the crack tip of 7050-T7451 aluminum alloy by Franc2D/L. Reprint from Ref. [37], Copyright 2011, with permission from Elsevier

$$\frac{\mathrm{d}a}{\mathrm{d}N} = \frac{C(1-f)^{n}\Delta K^{n}\left[1 - \frac{\Delta K_{th}}{\Delta K}\right]^{p}}{(1-R)^{n}\left[1 - \frac{\Delta K}{(1-R)K_{C}}\right]}$$

where C, , n, and p are empirically derived by curve fitting test data [39, 40], R is the stress ratio, f is the function of maximum applied stress and strain constraint factor, ΔK_{th} is the threshold value that is a function of R, K_{C} is a function of fracture toughness and specimen thickness, and ΔK is the range of the SIF.

The analysis results of the cracked configuration are used to compute the Mode I-type SIF which are used in a closed-form expression, and this expression determines the direction from the maximum circumferential stress around the crack tip. Propagation angle is also displayed to give the option of propagating the crack by an incremental distance along crack line during crack propagation.

2.3.3 Experimental Methods

2.3.3.1 Sample Preparation

7050-T7451 aluminum alloy was made into single-edged notched tensile (SENT) samples, as shown in Fig. 2.7. The alloy's chemical composition and mechanical properties are shown in Tables 2.1 and 2.2 separately.

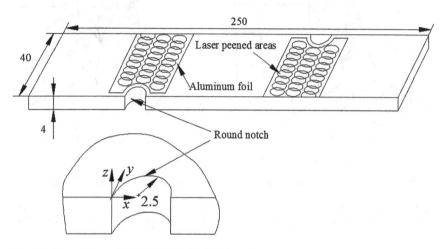

Fig. 2.7 Dimensions of the 7050-T7451 SENT specimen and laser shock processing route. Reprint from Ref. [37], Copyright 2011, with permission from Elsevier

Table 2.1 7050-T7451 chemical composition

Trademark	Si	Fe	Cu	Mn	Mg	Cr	Zr	Zn	Ti	Al
7050	≤0.12	≤0.15	2.0–2.6	≤0.10	1.9–2.6	≤0.04	0.08–0.15	5.7-6.7	≤0.06	Rest

Table 2.2 7050-T7451 mechanical properties

Trade mark	Supply condition	Thickness (Mm)	Sampling direct	No less than		
				Tensile stress (MPa)	Non-proportional extension strength $\sigma_{0.2}$ (MPa)	Elongation after break (%)
7050	T7451	60–76	L	503	434	9
			LT	503	434	8
			ST	469	407	3
		>76–102	L	496	427	9
			LT	496	427	6
			ST	469	400	3

Notes L is portrait line section; LT is landscape orientation line section; ST is thickness

2.3.3.2 Experimental Parameters

LSP was performed by high power Nd:glass laser implements to investigate the effect of the processing parameters on the residual stress distribution. Laser beam wavelength was 1.054 μm, pulse duration and power density were 20 ns and 2.91 GW/cm^2, respectively, and laser beam spot size was maintained at 8 mm. LSP experiment was conducted with a confined plasma configuration. A thin aluminum foil adhesive tape was used as the energy absorbing layer, and a water-tamping layer was used as the plasma confinement layer. The region with a hole was laser shocked, and nearly pure mechanical effects were induced, as shown in Fig. 2.8. The distributions of residual stress along the radial direction at the surface direction and depth direction were determined by using the X-350A X-ray diffraction technique. MTS880 material testing system was adopted for fatigue experiment in which tensile stress fatigue was adopted with load control. The stress ratio R was 0.1, alternating frequency was 20 Hz, σ_{max} was 460 Mpa, and the loading wave was sine wave.

2.3.4 Results and Discussions

2.3.4.1 Residual Stress Distributions

The laser-generated shock wave affects both the surface and side surface during its propagation. Residual stress distributions as a function of surface distance and depth are shown in Fig. 2.9. The contours of radial stress shown in Fig. 2.9a

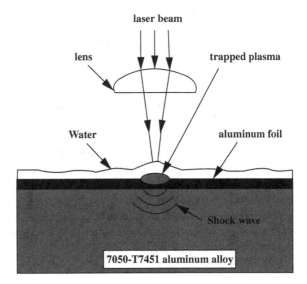

Fig. 2.8 Schematic of the laser shock processing principle. Reprint from Ref. [37], Copyright 2011, with permission from Elsevier

Fig. 2.9 Diagram of surface residual stresses of the 7050-T7451 aluminum alloy tensile SENT specimen induced by laser shock processing. **a** σ_{yy} in-surface and **b** σ_{xx} in-depth. Reprint from Ref. [37], Copyright 2011, with permission from Elsevier

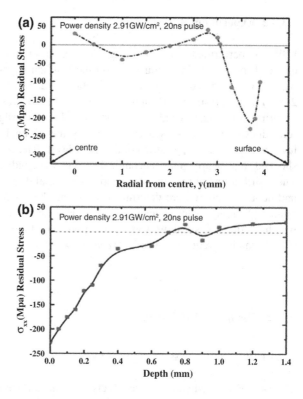

indicate that it is compressive with a maximum of 227 MPa at 20 ns with a 2.91 GW/cm^2 power density, and the residual radial stress distribution over the depth of the substrate is given as Fig. 2.9b. It is observed that the residual stresses are compressive. The maximum residual stress occurs at a depth of about 0.09 mm, and it is evident from the residual displacement at 20 ns that the axial displacement is confined to the beam impact region with a maximum of 1.0 mm. This pattern agrees well with the results reported in the literature [41].

2.3.4.2 Microscopic Analysis of Macroscopic Fracture

The laser-induced shock waves have a very high intensity which could reach several or even dozens of GPa, while their pulse widths are dozens of nanoseconds generally. Therefore, solid substances will generate some ultra-high strain rates, which are as much as 10^6–10^7 s^{-1} under the action of laser shock waves. Alloy materials usually have some special mechanical and physical characteristics under so high strain rates; their structures may generate dislocation, grains, and twins. And surface roughness, deformation, residual stress, and hardness of the material surface will be changed greatly [20, 21].

Typical macroscopic fracture morphology of the non-treated specimen (tested at 300 MPa) is shown in Fig. 2.10. Figure 2.10a shows the overall macroscopic appearance of the fracture, where internal longitudinal crack is seen in the central area of the specimen, and Fig. 2.10b shows the detail of the longitudinal crack. Several of these longitudinal cracks were formed near the central area of the specimen. As expected, the fatigue striation distance increased as the crack length increasing. The fatigue cracks are initiated from these cracks and grew until failure.

However, fatigue striation spacing in the LSP specimen (tested at 300 MPa) was narrower than those observed on the non-treated one, as shown in Fig. 2.11a, b. The distance between fatigue striations was narrow, and the number of striations was very large in the laser shock zones, which indicated that the crack expansion distance was small, and it means that LSP had an inhibition effect on the material fatigue crack initiation and expansions. The reduction in striation spacing indicated a slower

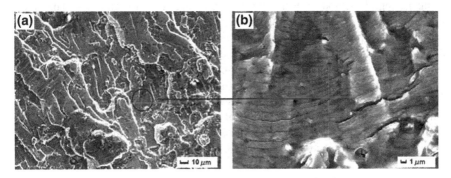

Fig. 2.10 Fractured surface of the non-LSP-treated 7050-T7451 aluminum alloy at an applied stress of 300 MPa, $R = 0.1$. **a** Macroscopic fracture appearance of a non-LSP-treated specimen and **b** detail of an internal crack across the center of the specimen. Reprint from Ref. [37], Copyright 2011, with permission from Elsevier

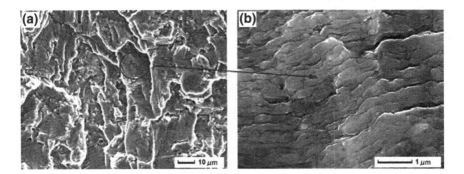

Fig. 2.11 Fractured surface of the LSP-treated 7050-T7451 aluminum alloy at an applied stress of 300 MPa, $R = 0.1$. **a** Macroscopic fracture appearance of a LSP-treated specimen and **b** detail of an internal crack across the center of the specimen. Reprint from Ref. [37], Copyright 2011, with permission from Elsevier

fatigue crack growth rates, which could be attributed to the compressive residual stress induced by LSP. As a result, the effective SIF that controls the fatigue crack growth in the LSP specimen is lower than that of the non-LSP case. The scatter in the spacing could be attributed primarily to the fact that striation formation is a highly localized event. The striation spacing is also dependent on both the SIF and metallurgical factors such as variations in the grain orientation [42].

2.3.4.3 Numerical Calculation of I-Type Cracks

Franc2D/L was applied in the stable-growth numerical calculation of I-type cracks on 7050-T7451 aluminum alloy. The elastic modulus E was defined as 71 GPa, Poisson ratio v was 0.345, and experimental load value DF was defined as 20 kN. After each step of crack growth, Franc2D/L calculated the SIF of each crack tip by using J-integral method to decide the next growing step of each crack. The crack length shows a monotonically increasing trend (Fig. 2.12). Nevertheless, a bifurcation will take place due to the nucleation of a vertical penny-shaped crack beneath the contact area as soon as the load is sufficiently increased. Therefore, the exact simulation of the propagation pattern requires a three-dimensional analysis, which is beyond the capability of the code unfortunately. The crack propagates following a sub-vertical pattern, and no bifurcations are observed. It is worth noting that the length is the only characteristic length of this structural scheme. Therefore, it is useful to plot the diagram of the non-dimensional SIF as a function of the crack length so as to emphasize the scale-invariant aspects of the problem.

Fig. 2.12 The relationship of the crack length and fatigue cycles along with the surface crack expanding curve with Franc2D/L. Reprint from Ref. [37], Copyright 2011, with permission from Elsevier

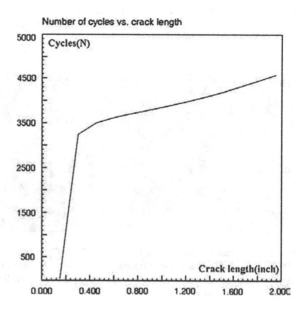

2.3.4.4 Improvement in Fatigue Life by LSP

The growth data of fatigue cracks were obtained based on the experimental data of the cracked sample by MTS880 fatigue experiment machine. The fatigue lives of specimens treated by LSP and un-treated specimens at the maximum applied stress of 460 Mpa are shown in Fig. 2.13. In an earlier study [43] of LSP-treated 7050-T7451 specimens, the expected large extension of fatigue life was not achieved. While in this study, LSP resulted in an improvement of 29 and 47 % in the number of fatigue cycles, after one and two laser shock layers (100 and 200 % coverage), respectively. The mean value in fatigue life of the specimens treated by LSP was higher than that of those untreated under the same maximum applied stress. This improvement is explained by the greater depth of the residual compressive stress fields induced by laser processing as comparing with non-treated sample. Further examination of these preexisting internal cracks was carried out to find an explanation for the fatigue lives of LSP specimens longer than expected. The number of fatigue cycles was compared to the experimental results when the crack length growing to 25 mm (1 in.). The computer-simulated fatigue circulating time was 38,509, while the fatigue experiment machine test value was 40,025. In the FNK model, numerical results are fairly good regarding the fatigue limit while they are conservative above it. This can be further demonstrated by plotting K_{res}, the residual value of K calculated from the residual stress field, as a function of distance from the leading edge. It is evident that Franc2D/L model provides good predictions for the crack growth circulating times.

A path following the initial crack growth is defined to make clear the residual stress field SIF. Figure 2.14 shows the Mode I-type SIF history for a single trace along the crack surface. The SIF varies along the crack front; thus, the SIF history and the predicted fatigue life would vary depending on the chosen path. Using the FNK fatigue crack growth model and the material parameters from Table 2.2, the

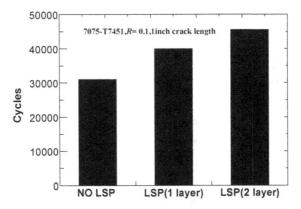

Fig. 2.13 Comparison of the fatigue crack expanding cycles of the 7050-T7451 aluminum alloy at the rate of $R = 0.1$ before and after LSP. Reprint from Ref. [37], Copyright 2011, with permission from Elsevier

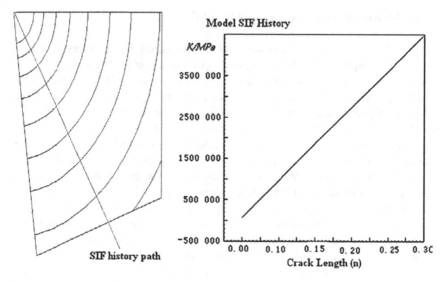

Fig. 2.14 Schematic of mode I-type SIF changing along the crack surface. Reprint from Ref. [37], Copyright 2011, with permission from Elsevier

number of predicted fatigue life starting from an initial surface crack is about 9200 cycles, as shown in Fig. 2.15. Due to the high levels of compressive residual stresses at the surface of the specimen, and analogous to the effect of LSP on crack growth, the cracks show considerable retardation upon reaching the treated region.

However, the curve of the crack length and the number of load cycle measurements did not show any considerable retardation of the fatigue cracks. This is believed to be due to the tensile core in the material that arises as a by-product of LSP. This tensile region is essential for the internal balance of forces in the unloaded component. Mahorter et al. [44] reported that the low cycle fatigue (LCF)

Fig. 2.15 Prediction of the fatigue life along with the crack length on 7050-T7451samples with Franc2D/L. Reprint from Ref. [37], Copyright 2011, with permission from Elsevier

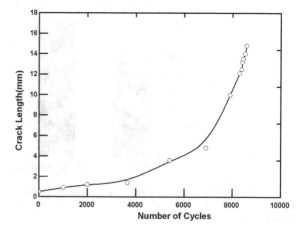

life of the disk ended when the crack in the bolt-hole reached 1/32 in. (0.79 mm), this corresponded to the life which was expressed as a 1/1000 probability of failure. The analyses presented here provide an estimate of remaining life once the LCF life is reached. Additional simulations could be performed to estimate the LCF life as well [36], but this is beyond the scope and purpose of this paper. It appears that the deflection point in the curve is barely noticeable, and hence, a sensitive measuring system requires to be detected.

2.3.5 Summary

The relationship of the 7050-T7451 aluminum alloy fatigue crack growth behaviors and surface SIF changing has been developed after nanoseconds LSP.

1. Compressive residual stress generated by LSP in crack surface reduces the tensile stress level in alternating load and decreases the effective driving force of fatigue crack growth. The reduction in striation spacing indicates a slower fatigue crack growth rates which would be attributed to the compressive residual stress (about to 250 Mpa) induced by LSP.
2. The effective SIF that controls the fatigue crack growth in the LSP specimen is lower than that of the non-LSP case. The more intense the residual stress produced by laser shock is, the smaller the effective SIF of crack surface becomes, resulting in the lower crack growing speeds.
3. FE analyses were carried out to determine the effective SIF at the crack tip. Reasonably, good agreements have been obtained from both the simulation and experiment results. The number of cycles to failure predicted numerically is lower than the experimental one. This difference is attributed mainly to an early stage of shear-dominated propagation that cannot be accounted for in the numerical model. All these illustrate the flexibility and interaction of Franc2D/L in simulating the process of crack growth with the residual stresses and SIF changing after LSP.

References

1. Prevey P et al (1998) Thermal residual stress relaxation and distortion in surface enhanced gas turbine engine components. In: Proceedings of the 17th heat treating society conference and exposition and the 1st international induction heat treating symposium. ASM Materials, Park, OH: pp 3–12
2. Paul SP et al (2001) The effect of low plasticity burnishing (LPB) on the HCF performance and FOD resistance of Ti-6Al-4 V. In: Proceedings of the 6th national turbine engine high cycle fatigue (HCF) conference, Jacksonville
3. Charles SM et al (2002) Laser shock processing and its effects on microstructure and properties of metal alloys: a review. Int J Fatigue 24:1021–1036

4. Amarchinta HK et al (2009) Material model validation for laser shock peening process simulation.Modell Simul Mater Sci Eng. http://iopscience.iop.org. Accessed 21 Jan 2013
5. Masmoudi N et al (1989) Influence of temperature and time on the stress relaxation process of shot peened IN 100 superalloys. Mater Tech (Paris) 77:29–36
6. Cao W et al (1994) Thermomechanical relaxation of residual stress in shot peened nickel base superalloy. Mater Sci Technol 10:947–954
7. Khadhraoui M et al (1997) Experimental investigations and modelling of relaxation behaviour of shot peening residual stresses at high temperature for nickel base superalloys. Mater Sci Technol 13:360–367
8. Cai DY et al. (2006). Precipitation and residual stress relaxation kinetics in shot-peened Inconel 718. J Mater Eng Perform 15(5):614–617
9. Zhong Z et al (2012) Thermal relaxation of residual stress in laser shock peened Ti-6Al-4 V alloy. Surf Coating Technol 206:4619–4627
10. Zhong Z et al (2011) A finite element study of thermal relaxation of residual stress in laser shock peened IN718 superalloy. Int J Impact Eng 38:590–596
11. Dennis JB et al (2009) A coupled creep plasticity model for residual stress relaxation of a shot peened Nickel-base superalloy. Mater Technol 131:75–79
12. Jeffrey JD et al (2009) Effects of material microstructure on blunt projectile penetration of a nickel-based super alloy. Int J Impact Eng 36:1027–1043
13. Marco DS et al (2003) Low and high velocity impact on Inconel 718 casting plates: ballistic limit and numerical correlation. Int J Impact Eng 28:849–876
14. Fabbro R, Pournier J, Ballard P (1990) Physical study of laser-produced plasma in confined geometry. J Appl Phys 68:775–784
15. Ren XD et al (2013) A finite element analysis of thermal relaxation of residual stress in laser shock processing Ni-based alloy GH4169. Mater Des 54(2014):708–711
16. Peyre P et al (1996) Laser shock processing of aluminium alloys. Application to high cycle fatigue behavior. Mater Sci Eng A 210:102–113
17. Evans A et al (2005) Relaxation of residual stress in shot peened Udimet 720 Li under high temperature isothermal fatigue. Int J Fatigue 27:1530–1534
18. Feng BX et al (2009) Residual stress field and thermal relaxation behavior of shot-peened TC4-DT titanium alloy. Mater Sci Eng A 512:105–108
19. Aghdam AB et al (2010) An FE analysis for assessing the effect of short-term exposure to elevated temperature on residual stresses around cold expanded fastener holes in aluminum alloy 7075-T6. Mater Des 31:500–507
20. Juijerm P, Altenberger I (2006) Residual stress relaxation of deep-rolled Al-Mg-Si-Cu alloy during cyclic loading at elevated temperatures. Scripta Mater 55:1111–1114
21. Xie LC et al (2011) Thermal relaxation of residual stresses in shot peened surface layer of (TiB + TiC)/Ti-6Al-4 V composite at elevated temperatures. Mater Sci Eng A 528:6478–6483
22. Ren XD et al (2013) Metallographic structure evolution of 6061-T651 aluminum alloy processed by laser shock peening: effect of tempering at the elevated temperatures. Surf Coat Technol 221:111–117
23. Almer JD et al (2000) The effects of residual macrostresses and microstresses on fatigue crack initiation. Mater Sci Eng A 284:268–279
24. Genel K et al (2000) Effect of ion nitriding on fatigue behaviour of AISI 4140 steel. Mater Sci Eng A 279:207–216
25. Berger MC, Gregory JK (1999) Residual stress relaxation in shot peened Timetal 21s. Mater Sci Eng A 263:200–204
26. Heinz A et al (2000) Recent development in aluminum alloys for aerospace applications. Mater Sci Eng A 280:102–107
27. Warren AW et al (2008) Massive parallel laser shock peening: simulation, analysis, and validation. Int J Fatigue 30:188–197
28. Caslaru R et al (2009) Fabrication and characterization of micro dent array produced by laser shock peening on aluminum surfaces. Trans NAMRI/SME 37:159–166

29. ArifAbulFazal M (2003) Numerical prediction of plastic deformation and residual stresses induced by laser shock processing. J Mater Process Technol 136:120–138

30. Ramsamooj DV (2003) Analytical prediction of short to long fatigue crack growth rate using small and large-scale yielding fracture mechanics. J Fatigue 25(9–11):923–933

31. Edgar HK (1998) Some aspects of fracture mechanics research during the last 25 years. Steel Res 69:206–213

32. Ray A, Patanker P (2001) Fatigue crack growth under variable amplitude loading: part I - model formulation in state space setting. Appl Math Modell 25:979–994

33. Ren XD et al (2009) Influence of compressive stress on stress intensity factor of hole-edge crack by high strain rate laser shock processing. Mater Des 30:3512–3517

34. Zhang YK et al (2009) Investigation of the stress intensity factor changing on the hole crack subject to laser shock processing. Mater Des 30:2769–2773

35. Carpinteri A, Pugno N (2006) Cracks and re-entrant corners in functionally graded materials. Eng Fract Mech 73:1279–1291

36. Barlow KW, Chandra R (2005) Fatigue crack propagation simulation in an aircraft engine fan blade attachment. Int J Fatigue 27:1661–1668

37. Ren XD et al (2011) Comparison of the simulation and experimental fatigue crack behaviors in the nanoseconds laser shocked aluminum alloy. Mater Des 32(2011):1138–1143

38. Qian J, Fatemi A (1996) Mixed mode fatigue crack growth: a literature survey. Eng Fract Mech 55(6):969–990

39. Forman RG et al (1994) Fatigue crack growth computer program NASA/FLAGRO version 2.0. Johnson Space Center, Houston (Texas): Rpt. # JSC-22267A

40. Newman JC Jr (1984) A crack opening stress equation for fatigue crack growth. Int J Fract 24: R131–R135

41. Gomez-Rosas G et al (2010) Laser shock processing of 6061-T6 Al alloy with 1064 nm and 532 nm wavelengths. Appl Surf Sci 256:5828–5831

42. Hatamleh Omar et al (2007) Laser and shot peening effects on fatigue crack growth in friction stir welded 7075-T7351 aluminum alloy joints. Int J Fatigue 29:421–434

43. Liu Q et al (2002) Internal cracking during surface treatment of 7050-T74 aluminium alloy using laser shock peening. Int Conf Struct Integrity Fract 25–28:177–182

44. Mahorter R et al (1985) Life prediction methodology for aircraft gas turbine engine disks. In: AIAA/SAE/ASME/ASEE 21st joint propulsion conference, Monterey, CA, pp 1–6

45. Sadananda K, Vasudevan AK (2005) Fatigue crack growth behavior of titanium alloys. Int J Fatigue 27:1255–1266

Chapter 3
Laser Shock Processing at Elevated Temperature

Abstract This chapter presents the influences on 00Cr12 alloy's mechanical properties at high temperatures, on metallographic structure evolution and dislocation configuration of 6061-T651 aluminum alloy at elevated temperature, and on ASTM: 410L00Cr12 microstructures and fatigue resistance in the temperature range 25–600 °C.

3.1 Introduction

Laser shock processing (LSP) is a new surface modification technology which generates high-strength impact wave and then introduces compressive residual stress of several hundred MPa by exposing metallic samples to high-power density (GW/cm^2), short-pulse (ns level) laser beam. The increasing compressive residual stress value and depth would significantly improve the fatigue life and corrosion resistance of metallic materials. For some cases, near-surface macroscopic compressive residual stresses and work-hardening states relax significantly or even completely under elevated temperature conditions, which will weaken the strengthening effect. The higher the dislocation density and internal energy of the material, the faster a given level of residual stress will relax at a fixed temperature. The higher the temperature, the faster the stress relaxation is. Thermal relaxation occurs in the initial period of aging, and as the time goes on, residual stress is approaching a fixed value.

LSP has a great effect on the surface topography and hardness of metallic materials. Generally speaking, surface micro-hardness of the materials has been significantly improved after the laser shock. However, micro-hardness changes at elevated temperatures. Apart from this, temperature has a significant relationship with the metallographic structure of the materials after LSP.

© Springer-Verlag Berlin Heidelberg 2015

X. Ren, *Laser Shocking Nano-Crystallization and High-Temperature Modification Technology*, DOI 10.1007/978-3-662-46444-1_3

3.2 Mechanical Properties and Residual Stresses Changing on 00Cr12 Alloy by Nanoseconds Laser Shock Processing at High Temperatures

Abstract Effects of laser shock processing (LSP) on 00Cr12 alloy's mechanical properties at high temperatures were analyzed by investigating the change of the fatigue properties, residual stress, and tensile properties. The results indicate that anti-fatigue life of material is enhanced greatly at room temperature after laser shock processing in which residual compressive stress is mechanically produced into the surface. The yield strength and the elasticity coefficient of 00Cr12 specimens are enhanced greatly after LSP; the cycle times are obviously longer at the elevated temperature, and the laser-shocked samples exhibit lower plastic strain amplitudes compared with the un-treated ones. The compressive residual stress produced by LSP, which makes the 00Cr12 alloy's yield strength and tensile strength increase, still played an important role in the high-temperature tensile test. The tensile fracture showed that the break position and expand direction were changed because of the residual compressive stress existing, and the residual compressive stress was greatly released with the increasing temperature. The results demonstrated that the cycle, stress amplitude, and temperature-dependent relaxation of compressive residual stresses were more pronounced than the decrease of near-surface work strengthening.

3.2.1 Introduction

00Cr12 steel is a kind of heat-resistant steel with good thermostable susceptibility and organizational stability, which are usually used in the elevated temperature condition. Many materials, such as 316L steel [1, 2], 2.25Cr1Mo steel [3], Ti6Al4 V [4], 304 Steel [5], and 1.25Cr0.5 M Steel [6], have been deeply studied about the fatigue properties at elevated temperature. Few researchers have investigated the fatigue properties of heat-resistant steel materials from the room temperature to 600 °C, especially the influence of the elevated temperature has not been carried on through careful experimentation and fundamental research [7].

The wear-resisting and the fatigue resistance capability of the heat-resistant steel surface would be improved by the superficial strengthening technology. Thus, its fretting fatigue resistance is strengthened [8], and its fatigue life is enhanced [9]. LSP has been recognized as the most effective method to enhance the fretting fatigue resistance of the material in room temperature condition [10, 11]. In the elevated temperature environment, the residual compressive stress of laser-strengthened layer would relax [12], and the strengthened effect would decrease [13], but the destruction and the damage mechanisms of components in the elevated temperature environment are complicated [14, 15]. Therefore, the study on the elevated temperature rule after LSP is of practical value in engineering. With respect to the continuity information between this study and the available literature, the fatigue test and analysis on the 00Cr12 steel after LSP from room temperature to 600 °C are performed in this study, and we aim to obtain the elevated temperature fatigue behaviors of the material after LSP.

3.2.2 LSP on Fatigue Properties at Room Temperature

Chemical compositions of 00Cr12 are shown in Table 3.1, and schematic diagrams of the standard tensile fatigue specimen after LSP are shown in Fig. 3.1.

LSP was performed on 00Cr12 specimens by Q-switched high-power neo-dymium-doped glass laser with a wavelength of 1.054 μm, pulse duration of 20 ns, energy of 18 J, and the laser power density of 3.29 GW/cm^2. Water was used as the transparent overlay, and aluminum foil with 0.1 mm thickness was used as the absorbent coating. After LSP, pulling test was performed on the specimens by GPS200 high-frequency fatigue tester at 25 °C. The electronic universal testing machine was used to pull the specimen until broken. 00Cr12 mechanical properties at room temperature are shown in Table 3.2.

The high-frequency fatigue test was performed before and after LSP at a stress level of 270 MPa, and two groups of fatigue lives were obtained. An accurate

Table 3.1 Chemical compositions of 00Cr12 alloy

Brand	C	Si	Mn	S	P	Cr	Ni
00Cr12	≤0.030	≤1.00	≤1.00	≤0.030	≤0.035	11.00–13.00	≤0.60

Fig. 3.1 Dimensions and schematic diagrams of the 00Cr12 tensile fatigue specimen. Reprint from Ref. [16], Copyright 2010, with permission from Elsevier

Table 3.2 Mechanical properties of 00Cr12 alloy

Brand	Elastic modulus, E (MPa)	Tensile strength, σ$_b$ (MPa)	Yield strength, σ$_{0.2}$ (MPa)	Cross-sectional shrinkage (%)	Elongation, δ$_{10}$ (%)
00Cr12	0.226	895	275	19	8

estimate can be obtained by inferential error bars such as the standard error (SE). SE is defined as

$$SE = \sqrt{\frac{\sum(x - M)^2}{n(n - 1)}}$$

where x refers to the individual data points, M is the mean, and $\sum$(sigma) means adding to find the sum, for all the n data points.

SE is, roughly, a measure of how variable the mean will be, if you repeat the whole study many times. The SE varies inversely with the square root of n, so the more often an experiment is repeated, or more samples are measured, the smaller the SE becomes. The length of each arm of the SE bars is shown in Fig. 3.2, which indicates that the fatigue life of metallic material is enhanced largely after LSP at room temperature compared with the virgin material.

The beneficial effect on fatigue life at room temperatures from LSP can be ascribed to the creation of significant near-surface compressive residual stresses and a near-surface work-hardened layer.

The fatigue safety lives were calculated by the one-sided allowance factor method in the confidence of 95 % and reliability of 99 %, and the results are shown in Fig. 3.3 with the accurate estimate by inferential SE bars.

The results show that the anti-fatigue life of specimens is enhanced significantly after LSP, and their stress fatigue life and fatigue safety life are improved 62 and 75 % more than the un-treated samples, respectively. The reinforced layer is formed in the metallic material induced by high-pressure shock waves [17, 18], the residual compressive stress is very large, superficial hardness is very high in the surface [19], and the micro-characteristics of its inner layer are crowded dislocations and refinement crystal grains.

Fig. 3.2 Enhancement in fatigue life of different specimens before and after LSP, with the accurate estimate by inferential SE bars. Reprint from Ref. [16], Copyright 2010, with permission from Elsevier

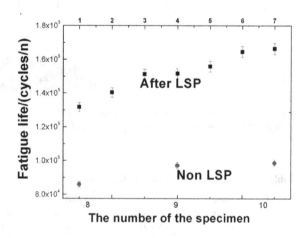

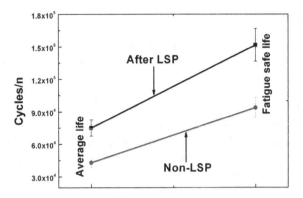

Fig. 3.3 Comparison of the changing of mean average life and fatigue safety life before and after LSP, with the accurate estimate by inferential SE bars. Reprint from Ref. [16], Copyright 2010, with permission from Elsevier

3.2.3 LSP on Fatigue Properties at Elevated Temperature

The fatigue behaviors of 00Cr12 at elevated temperatures involve many factors. Tensile and cycle fatigue tests were performed on 00Cr12 steel specimens before and after LSP from room temperature to 600 °C. The smooth round bar specimen was used in this experiment, as shown in Fig. 3.4.

MTS809 Axial/Torsion Test System was used to test tensile and cycle fatigue behaviors, and a three-zone resistance-type furnace within a variation of ±1 °C at steady state was used for temperature control. A high-temperature extensometer (MTS model No. 632-13F-20, gauge length 25 mm) was used to measure the strain and control the strain signal. Tensile and cycle fatigue tests were conducted in air by maintaining a constant cross-head speed of 3 mm/min, employing a sine waveform, at room temperature, 400, 500, and 600 °C, respectively. The strain amplitude of 0.2 % was employed for all cycle fatigue tests.

Fig. 3.4 Dimensions and schematic diagrams of the 00Cr12 smooth round bar specimen of untreated and LSP. Reprint from Ref. [16], Copyright 2010, with permission from Elsevier

3.2.3.1 Yield Strengths

The yield strengths of 00Cr12 before and after LSP at different temperatures are shown in Fig. 3.5, and their elasticity modulus is shown in Fig. 3.6 with the accurate estimate by inferential SE bars.

The results show that, after LSP, the 00Cr12 specimen's yield strength and elasticity modulus are enhanced greatly with the increase in temperature. The results also indicate that, at elevated temperature, although residual compressive stress relaxation occurs and dislocations slip easier, the residual compressive stress would balance the tensile stress in the tensile test due to the shorter test duration and high dislocation density after LSP. Compared with un-LSP, laser-induced shock waves have a very high intensity which would reach several or even dozens of GPa, while their pulse widths are generally dozens of nanoseconds. Therefore, solid substances will generate some ultra-high strain rates which are as much as 10^6–

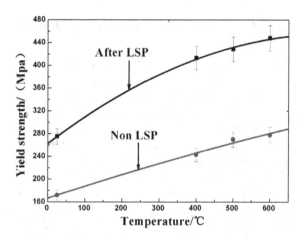

Fig. 3.5 Comparison of yield strengths of 00Cr12 steel before and after LSP at different temperatures. Reprint from Ref. [16], Copyright 2010, with permission from Elsevier

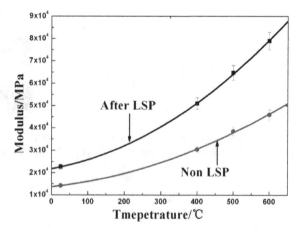

Fig. 3.6 Comparison of elasticity modulus of 00Cr12 steel before and after LSP at different temperatures. Reprint from Ref. [16], Copyright 2010, with permission from Elsevier

10^7 s^{-1} under the action of laser shock waves. Compressive residual stresses are reduced with the increasing tensile strain and tend to disappear completely.

3.2.3.2 Effects of LSP on Cycle Times

Different load levels have different effects on cycle times [13]. The temperature also has a great influence on the heat-resistant steel 00Cr12 life after LSP, as shown in Fig. 3.7.

With the increase in temperature, the cycle times are significantly reduced. It shows that the higher the levels of the load curve are, the larger the slope are, indicating that the temperature has a greater effect on the cycle life of high load levels. At the same load level, the curve slopes at elevated temperature segments are larger than those at the low temperature. The slope slows down at the range of 400–500 °C, but the curve sloping is much faster in the range of 500–600 °C, indicating that the higher the temperature is, the faster the life cycle reduces, and the results imply that LSP has the greatest effect on the 00Cr12 at less than 500 °C. This is mainly high temperature oxidation of materials during the fatigue deformation. Obviously, the higher the test temperature and load level are, the more significant the oxidative damage of the sample is and the lower the fatigue life of specimens is.

3.2.3.3 Effects of LSP on Axial Strain

The relationship between the fatigue life and the axial strain of specimens before and after LSP was explored at 400 and 600 °C. The area of the stress–strain loop decreases with increasing number of cycles at 400 °C, while the slope increases, as shown in Fig. 3.8.

Fig. 3.7 Relational curves of cycle times and test temperatures at the strain amplitude of 0.3, 0.4, 0.5, 0.6, and 0.7. Reprint from Ref. [16], Copyright 2010, with permission from Elsevier

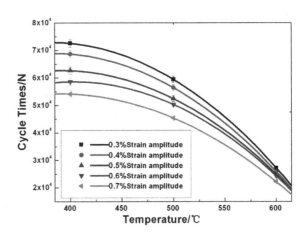

Fig. 3.8 The relationship
between the fatigue life and
axial strain at 400 °C of untreated and LSP. Reprint from
Ref. [16], Copyright 2010,
with permission from Elsevier

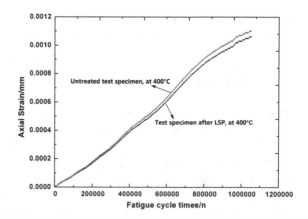

It is not surprising that further cycling beyond the initial cycle appears to cause very little further relaxation at 400 °C. Therefore, the material cyclically hardens. This is not so apparent at 400 °C, where the average load or the exposure times may not be sufficient to activate creep mechanisms. Conversely, at higher temperatures, the area of the hysteresis loop does not decrease, while the gradient decreases slightly, suggesting that the material does not strain and harden above 600 °C. As a result, further residual stress relaxation after the initial cycle is apparent at high temperatures, as shown in Fig. 3.9.

These results indicate that LSP reduces the expansion rate of fatigue, so the fatigue life of materials is increased, while raising the temperature would increase the expansion rate of fatigue, thereby reducing the fatigue life of materials. This phenomenon is due to the significant plastic strain induced during the first fatigue cycle. Fatigue also causes a greater reduction of the integer width relative to the solely thermal case, indicating that high-temperature isothermal relaxation at 600 °C is perhaps driven by creep.

Fig. 3.9 The relationship
between the fatigue life and
axial strain at 600 °C of untreated and LSP. Reprint from
Ref. [16], Copyright 2010,
with permission from Elsevier

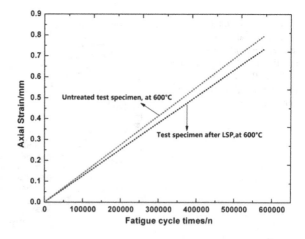

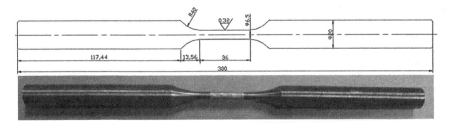

Fig. 3.10 Dimensions and schematic diagrams of the standard tensile fatigue specimen of 00Cr12 alloy after LSP. Reprint from Ref. [20], Copyright 2011, with permission from Elsevier

3.2.4 LSP on Residual Stresses at Elevated Temperature

00Cr12 specimens, as shown in Fig. 3.10, were performed by Q-switched high-power neodymium-doped glass laser with a wavelength of 1.054 μm, pulse duration of 20 ns, and power density of 1.54 GW/cm^2 in the strong laser laboratory of Jiangsu University. Large-diameter quarter-wave plate and high-strength large-diameter film polarizer were adopted. The laser beam spot size was maintained at a diameter of 6 mm, and the overlapping rate of the laser spot was 50 %. Water was used as the transparent overlay, and aluminum foil with 0.1 mm thickness was used as the absorbent coating.

After LSP, pulling test was performed on specimens by MTS809 Axial/Torsional Test System at different temperatures, and a three-zone resistance-type furnace, which could control temperature within a variation of ±1 °C (±3 °C for room temperature) at steady state, was used for temperature control. A high-temperature extensometer (MTS model No. 632-13F-20, gauge length 25 mm) was used to measure the strain and control the strain signal. Each specimen was heated at constant temperature for 5 min and then pulled until broken with a speed of 3 mm/min, and the value of yield strength by 0.2 % strains was recorded. The residual stresses in the laser impact zone of the specimen were determined by using the X-350A X-ray stress tester with sin$^2\psi$ method, and the tensile fracture was observed with a JEOL-6700 scanning electron microscope (SEM).

3.2.4.1 Residual Stress

Generally, the components should avoid high temperature that would release the residual compressive stress after LSP, which would weaken the effect of LSP. According to AMS 2546 "Laser shock processing standard (2004.8)," the heating temperatures of stainless steel should not exceed 750 °F (399 °C) [21], but there is no clear specification to the heat-resistant steel. Figure 3.11 shows the residual stress relaxation of 00Cr12 alloy at different temperatures.

Surface average compressive residual stress of 00Cr12 alloy after two LSP impacts is about −382 MPa. It could be found that the increase in amplitude of

Fig. 3.11 Redistribution diagram of the residual stress relaxation of 00Cr12 alloy at different temperatures. Reprint from Ref. [20], Copyright 2011, with permission from Elsevier

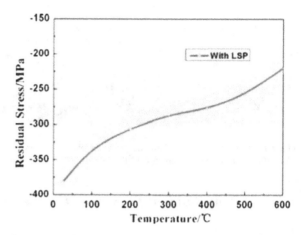

stress relaxation was large when the temperature was lower than 200 °C, and the increase in amplitude tended to slow down between 200 and 500 °C, while increasing after 500 °C. When the temperature surpassed to 600 °C, the compressive residual stress reduced in a large scale. The average residual compressive stress dropped to −212.5 MPa, decreasing about 64 % at the temperature of 600 °C. It could be concluded that the working temperature of 00Cr12 alloy should not exceed 600 °C after LSP. And the low-temperature tempering treatment could be carried out before the tensile test in order to avoid the residual compressive stress significant relaxation. Residual compressive stress plays a dominant role in enhancing the anti-fatigue properties of the specimen, and the anti-fatigue properties would be significantly reduced in high-temperature conditions after LSP. Monotonic and cyclic relaxations are the reasons why residual stresses are not stable at high temperature. In addition, compressive stresses are not stable in the elevated temperatures and plastic deformations of service conditions. However, the residual compressive stress does not have enough time to release during the high-temperature tensile process, so it will balance the tension in the process of tensile stress [22, 23]. Therefore, LSP will be beneficial to enhance the yield strength and the tensile strength of 00Cr12 alloy.

3.2.4.2 Stress and Strain

The tensile stress–strain curves of 00Cr12 alloy at 400 and 600 °C are shown in Fig. 3.12a, b, which indicated that the testing temperature had no significant effect on the elongation of the 00Cr12 alloy. Even at the testing temperature of 600 °C (Fig. 3.12b), the fracture elongation was a little reduced for the 00Cr12 alloy.

It was found that the shape of the stress–strain curves changes with increasing temperature. Figure 3.13a shows that the elastic modulus and the cross-sectional shrinkage rate increase, obviously, while the yield strength and tensile strength

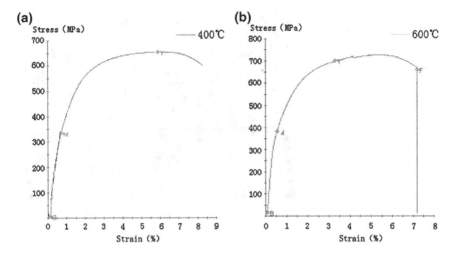

Fig. 3.12 Tensile stress–strain curves of 00Cr12 specimens at different temperatures. **a** 400 °C; **b** 600 °C. Reprint from Ref. [20], Copyright 2011, with permission from Elsevier

increase slightly with the increasing temperature after LSP. This could be ascribed to the surface micro-hardness increasing after LSP [24, 25]. The plasticity of 00Cr12 alloy is enhanced at high temperatures, which could be found in Fig. 3.13b.

The thermal activation increases with the temperature elevating, and the dislocation slips become easier. But the tensile process is so short (within the ns progress) that the residual compressive stresses have little time to release. The residual compressive stress would counteract the tensile stress partially in the stretch process which is equivalent to reduce the load of the material, so the tensile strength is improved.

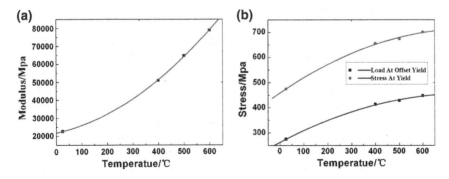

Fig. 3.13 Schematic of mechanical properties changing at different temperatures. The relationship of the **a** elastic modulus and the temperatures, and **b** stresses and the temperatures. Reprint from Ref. [20], Copyright 2011, with permission from Elsevier

Fig. 3.14 Fracture morphology of specimens at the temperatures of 25, 400, 500, and 600 °C, respectively. Reprint from Ref. [20], Copyright 2011, with permission from Elsevier

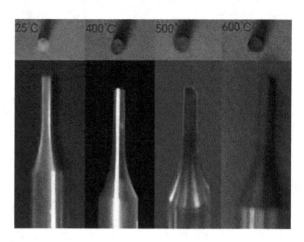

3.2.4.3 Fracture Morphology

Figure 3.14 displays the fracture morphology of specimens at different temperatures. It could be clearly found that the fracture morphology of specimens is greatly different from each other at the temperatures of 25, 400, 500, and 600 °C, respectively.

Compared to room temperature, the sample surface color is slightly deeper than that at the temperature of 400 °C, and the fracture face exhibits gray color and slight metallic luster. While at the temperatures of 500 and 600 °C, the specimen surface color is obviously deepened and presents grayish black, and the fracture face color is blue black without metallic luster. The "neck shrinks" phenomenon is not very obvious when the sample approaches to be broken, and it could be obtained that 00Cr12 is one kind of ferritic heat-resistant steel which has high structure stability and high-temperature oxidation resistance working long term at high temperatures.

3.2.4.4 Microstructure Morphology

LSP also results in fatigue properties changing at high temperatures. It could be found that the precipitates and dislocation are entangled after LSP even at high temperature. The morphologies of 00Cr12 alloy surface at different temperatures are shown in Fig. 3.15. Figure 3.15a shows that the fracture cracks originate at the center region at the room temperature and then extend radiated to the edge in the form of radiation with a few pores, showing the intense planar sliding. Figure 3.15b shows the SEM of the middle region at 25 °C, which has formed typical dimples with a few inclusions, and composition analysis indicates that the inclusions are manganese sulfide and some carbon. Figure 3.15c shows that the sulfide inclusion which is bigger in size and distributed in clusters is crack initiation. Cracks occur easily in those defects, and expand along its cleavage plane and produce axial

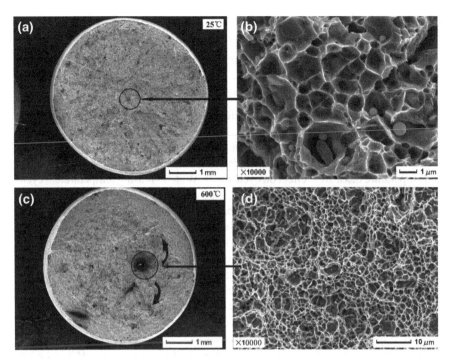

Fig. 3.15 SEM of 00Cr12 alloy under different temperatures. **a** Fracture cracks originate at the center region at the room temperature. **b** SEM of the middle region at 25 °C which has formed typical dimples. **c** The crack initiation at the 600 °C. **d** SEM of the middle region at 600 °C. Reprint from Ref. [20], Copyright 2011, with permission from Elsevier

cracks, which indicate that residual compressive stresses do not release completely during the high-temperature tensile test, indicating that residual compressive stresses still influence the expansion path of cracks in early stage [26–28]. It could also be found that the dimple is more homogeneous at 600 °C than at 25 °C, as shown in Fig. 3.15d.

At the room temperature, the cleavage plane and tearing ridge, which were usually taken as typical brittle fracture features, could be seen on the tensile fracture surfaces of the 00Cr12 alloy. At the elevated temperatures, however, the tensile fracture surface of the 00Cr12 alloy was composed of the cleavage plane, tearing ridge, and dimple with small size. It implied that the 00Cr12 alloy showed a mixed ductile and brittle fracture feature. It is important to note that this beneficial effect of the mechanical surface treatment is not restricted solely to ambient temperatures, but it is also clear in evidence at 600 °C. This reveals that precipitates are more likely to nucleate around the dislocations during dynamic aging in LSP.

3.2.5 Conclusions

LSP is considered to be one of the most promising techniques in terms of its ability to induce compressive residual stresses, which improves the mechanical performance of the materials. The obtained results are as follows:

1. At room temperature, the fatigue life of material is greatly improved after LSP. In confidence of 95 % and reliability of 99 % with the accurate estimate by inferential SE bars, the fatigue life and fatigue safety life are improved 62 and 75 % more than the un-treated samples, respectively.
2. Temperature has a great effect on cycle times. With increasing temperatures, cycle times are significantly reduced. Temperature also has a greater effect on the cycle times with a higher load. The higher the test temperature and load level are, the more significant oxidative damage of the sample is, and the lower the fatigue life of specimens is. The yield strength and the elasticity coefficient of the 00Cr12 are enhanced greatly after LSP even at elevated temperature up to 600 °C.
3. LSP reduces the expansion rate of fatigue of materials and increases the fatigue life of materials; the area of the hysteresis loop does not decrease, while the gradient decreases slightly after LSP at higher temperatures. This also implies that the slope of the stress–strain curve decreases with the amount of tensile work at a given stress.
4. LSP introduces a compressive residual stress field at the surface of 00Cr12 alloy (up to −382 MPa). The residual compressive stress will be relaxed under high temperature. Monotonic and cyclic relaxations are the reasons why residual stresses are not stable at high temperature. The average residual compressive stress dropped to −212.5 MPa, decreasing about 64 % at the temperature of 600 °C.
5. LSP successfully enhances the yield strength and tensile strength of the 00Cr12 alloy. The residual compressive stress would counterbalance the tensile stress partially in the stretch process which is equivalent to reduce the load of the material, so the tensile strength is improved.
6. Residual compressive stresses induced by LSP still influence the expansion path of cracks in early stage at the high-temperature tensile test, and the results imply that LSP has the greatest effect on the 00Cr12 at less than 600 °C.

3.3 High-Temperature Mechanical Properties and Surface Fatigue Behavior Improving of Steel Alloy Via Laser Shock Peening

Abstract Laser shock peening (LSP) was carried out to reveal the effects on ASTM: 410L 00Cr12 microstructures and fatigue resistance in the temperature range 25–600 °C. The new conception of the pinning effect was proposed to explain the improvements at the high temperature. Residual stress was measured by X-ray diffraction (XRD) with $\sin^2\psi$ method, and a high-temperature extensometer was utilized to measure the strain and control the strain signal. The grain and precipitated phase evolutionary process were observed by scanning electron microscopy. These results show that a deep layer of compressive residual stress is developed by LSP, and ultimately, the isothermal stress-controlled fatigue behavior is enhanced significantly. The formation of high-density dislocation structure and the pinning effect at the high temperature induce a stronger surface, lower residual stress relaxation, and more stable dislocation arrangement. The results have profound guiding significance for the fatigue-strengthening mechanism of components at the elevated temperature.

3.3.1 Introduction

Many components work in high-temperature conditions, and the thermomechanical stability of the microstructure is related to the fatigue behavior of the treated materials at elevated temperatures. Conventional fatigue-strengthening mechanism generally produces a surface layer with high compressive residual stress and work hardening [29, 30], such as shot peening and deep rolling. Moreover, these traditional fatigue-strengthening mechanisms are only effective at the room temperature [31]. Furthermore, it was found that work-hardening and residual stress may be reduced significantly after cyclic loading [32], especially at high temperatures.

Half of the initial compressive stress may disappear in 10 min even at moderate engine temperature [31]. However, the improvement in the fatigue lifetime of mechanical surface treatments is known to depend mostly on the depth distribution, stability of the induced residual stresses, and work hardening in the near-surface regions [33].

LSP is an established mechanical surface treatment technology, which is generally utilized to improve the metallic surface [34, 35]. Compared with conventional surface mechanical treatment methods [36, 37], LSP could change near-surface microstructure and residual stress state, and eventually make material present a higher thermal stability. In addition, the treated component in LSP has a lower surface roughness, which reduces cracking initiation at the surface [38]. Therefore, material surface after LSP would be expected to be suitable under the high-temperature fatigue conditions.

According to Aerospace Material Specification (AMS) 2546 "Laser peening standard (2004.8)" [39], the working temperature of stainless steel should not exceed 399 °C after LSP, but there is no clear specification to the heat-resistant steel. Moreover, few researchers have investigated whether LSP would play the same effect at high temperatures, and how is about the relationship that fatigue behavior and stress stability of heat-resistant steel after LSP at high temperature.

In this section, the isothermal fatigue tests were carried out to investigate the fatigue behavior of ASTM: 410L (Standard Test Method for Wear Layer Thickness of Resilient Floor Coverings by Optical Measurement) 00Cr12 steel after LSP in the temperature range of 25–600 °C. The effect of the LSP was also investigated for improving the fatigue behavior of ASTM: 410L 00Cr12 at high temperature.

3.3.2 Experiments and Analysis

The material utilized in this study is the ASTM: 410L 00Cr12 steel, which is one kind of heat-resistant steel with good thermostable susceptibility and organizational stability and usually utilized at the elevated temperatures. The composition of ASTM: 410L00Cr12 alloy is given in Table 3.3, and the dimensions and schematic of standard stretch fatigue ASTM: 410L 00Cr12 specimen are shown in Fig. 3.16.

The LSP was performed using a nanosecond Nd:YAG laser with the pulse duration of 20 ns and laser power density of 3.29 GW/cm^2, and the spot-size diameter was approximately 6 mm. Surrounding specimen surfaces were also treated under the same conditions.

Residual stress was measured at different positions across the LSP regions by using a standard XRD technique with the sin$^2\psi$ method. Depth profiles were obtained by successive electrolytic removal of the material. The residual stress relaxation by sample cyclic loading was investigated using MTS880 closed-loop universal testing machine. The load was cycled between 0 and 350 MPa upper limits (equivalent to 85 % of yield strength), at a stress ratio $R = 0.1$, starting from low to high. A high-temperature extensometer (MTS model No. 632-13F-20, gauge length 25 mm) was utilized to measure the strain and control the strain signal. Isothermal fatigue experiments were performed under load control intension–compression on a standard

Table 3.3 Chemical compositions of 00Cr12 alloy

Composition	C	Si	Mn	S	P	Cr	Ni
Percent (wt%)	≤0.030	≤1.00	≤1.00	≤0.030	≤0.035	11.00–13.00	≤0.60

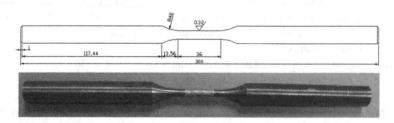

Fig. 3.16 Dimensions and schematic of standard stretches fatigue 00Cr12 specimen. Reprint from Ref. [40], Copyright 2014, with permission from Elsevier

testing machine with a load ratio of $R = 0.5$, a frequency of 40 Hz, and temperatures of 25–600 °C. The cylindrical samples were heated with a radiant heating. In order to minimize thermal gradients, 10 min earlier than the actual fatigue tests began to heat the smooth cylindrical specimens. Scanning electron microscopy (SEM) was employed to observe the grain and precipitated the phase evolutionary process. All of these measurements are intended to give an indication of the local fatigue behavior after LSP.

3.3.3 Results and Discussions

3.3.3.1 Residual Stress

LSP induced compressive residual stress of about 380 MPa at the ASTM: 410L 00Cr12 surfaces and 400 MPa in the subsurface maximum in a depth of approximately 0.1 mm, as shown in Fig. 3.17.

The compressive residual stresses play an important role in the retard subsequent crack propagation [41]. The variation of the diffraction peak width across the laser-shocked sample suggests that the sample has yielded significant plastic deformation to a depth of around 1 mm. And even the residual stress has also been changed from compressive stress to tensile stress due to the plastic-affected depth. The higher plastic-affected depth leads to a higher resistance to crack propagation during fatigue cyclic loading. According to Nikitin and Altenberger [42, 43], plastic-affected depth would significantly affect the fatigue behavior of materials, especially in push-pull loading fatigue test utilized in this study.

The physical process of the LSP is rather complex due to the multiphysical phenomena [44]. A better understanding of the mechanism of high-temperature residual stress relaxation is beneficial for developing a physics-based relaxation model. LSP-induced residual stress would release in high-temperature process for further plastic

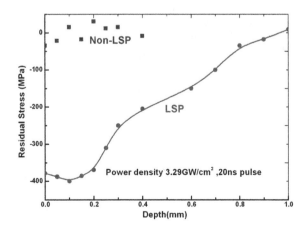

Fig. 3.17 X-ray diffraction measurements revealed that LSP produced compressive residual stresses in the near-surface regions. Reprint from Ref. [40], Copyright 2014, with permission from Elsevier

deformation even under relatively low cyclic loading. The stress relaxation process could be expressed by the Zener–Wert–Avrami function as [45, 46]

$$\frac{\sigma_t}{\sigma_0} = \exp[-(At)^m] \qquad (3.1)$$

$$A = B\exp(-\frac{Q}{RT}) \qquad (3.2)$$

where σ_t is the surface residual stress in high-temperature process, σ_0 is the surface residual stress after LSP, m is a numerical parameter dependent on the dominant relaxation mechanism, t is the aging time, A is a function dependent on the material and temperature, B is a constant, and Q is the activation enthalpy for the relaxation process. R is the Boltzmann constant, and T is the aging temperature.

This model incorporates the residual stress relaxation of heat-resistant steel in the temperature-enhancing process. The residual stress relaxation rate is proportional to the applied stress amplitude [47] and the number of stress cycles. As shown in Fig. 3.18, with the increase in holding time, the compressive flow strength decreases for a given plastic offset.

In addition, residual stress relaxation increases gradually with the increase in temperature. After the sample was incubated for 15 h at the temperatures of 400 and 500 °C, the residual stresses would release by 50 and 62 %, respectively. This fully illustrates that the residual stresses are not stable at high temperature. Moreover, compressive stresses are also not stable under the elevated temperature and plastic deformation conditions. Thermal exposure to service temperatures would reduce residual stresses by more than 50 %, the underlying mechanism is very similar to the well-known stress relaxation observed for creep under constant strain, and this pattern is coherent with the results reported in the literature [48].

Fig. 3.18 Surface residual stress relaxation after different insulation hours at the temperatures of 400 and 500 °C. Reprint from Ref. [40], Copyright 2014, with permission from Elsevier

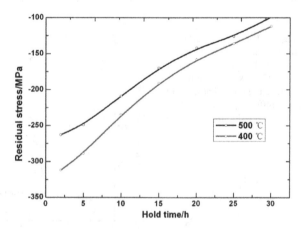

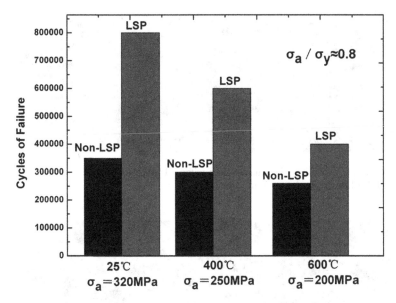

Fig. 3.19 Enhancement in fatigue lifetimes following LSP for test temperatures of 25, 400, and 600 °C. Reprint from Ref. [40], Copyright 2014, with permission from Elsevier

3.3.3.2 Fatigue Cycle Life

Figure 3.19 shows the fatigue life of samples before and after laser shocked, which clearly indicates that the fatigue life of ASTM: 410L 00Cr12 is significantly enhanced by LSP in the temperature range of 25–600 °C.

The fatigue life is enhanced about 128 % in the low cycle regime of 10^5 cycles at 25 °C. Already at 400 °C, the fatigue life has been enhanced by 102 %. Moreover, at the highest temperature of 600 °C, the fatigue life has also been enhanced by 53.8 %, although reduced as compared to that at 400 °C. These results may come from the creation of near-surface compressive residual stresses, and a near-surface work-hardened layer which is stable during thermal exposure or fatigue cycling at temperatures up to or exceeding 600 °C. The effect of this work hardened layer is to reduce the plastic strain amplitude, which acts to lessen the driving force for fatigue damage. In addition, the yield strength and ultimate tensile strength are also increased after LSP. Between 25 and 600 °C, the laser-shocked state is always exhibiting lower plastic strain amplitudes than the untreated state.

3.3.3.3 Fracture Surface

Figure 3.20 shows SEM micrographs of untreated ASTM: 410L 00Cr12 at 400 °C. The untreated state exhibits a fully austenitic microstructure and some dislocation tangles with low density. Figure 3.21 represents the near-surface microstructures after LSP at 400 °C.

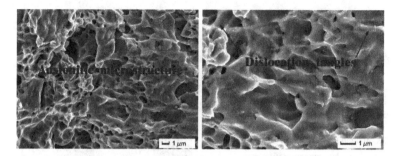

Fig. 3.20 Steady expansion of fracture surface without LSP at 400 °C. Reprint from Ref. [40], Copyright 2014, with permission from Elsevier

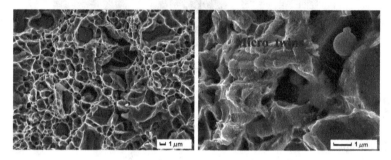

Fig. 3.21 Steady expansion of fracture surface after LSP at 400 °C. Reprint from Ref. [40], Copyright 2014, with permission from Elsevier

A layer of deformation-induced martensite and micro-twins are presented in near-surface regions. However, there were neither any nano-crystalline layers nor pronounced deformation-induced martensite at the surface. Relatively, the higher strain rate would accelerate the generation of twin crystal [49]. LSP exhibits a strain rate of about 10^5 s^{-1}, which is significantly higher than that of the deep rolling of about 10^{-1} s^{-1}. In terms of strain rates, the generation rate of twin crystal should be higher by LSP as compared to deep rolling [32].

Mechanical treated surface is generally considered to be able to enhance S/N fatigue lifetimes primarily by inhibiting the initiation of cracks. LSP also leads to the lower initial fatigue crack growth rate at high temperature. Figure 3.22 shows SEM images of the striations and the local crack growth rates at 400 and 600 °C, estimated from their spacing on the fracture surfaces of un-LSP and LSP, at stress amplitudes of 200 and 25 MPa.

These results indicate that LSP has an additional positive effect on fatigue properties by lowering the initial fatigue crack growth rates, typically by a factor of half, as compared to the corresponding behavior of the untreated sample. It is important to note this beneficial effect that the mechanical surface treatment is not restricted solely to ambient temperatures, but is also clear at 600 °C. In addition, the

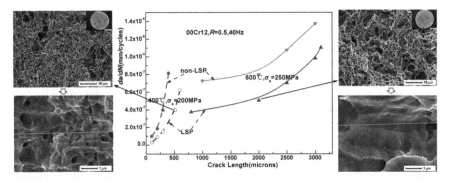

Fig. 3.22 Local fatigue crack growth rates at 400 and 600 °C, estimated from striation spacing measurements on fracture surfaces of un-treated and LSP. Reprint from Ref. [40], Copyright 2014, with permission from Elsevier

precipitated phases and dislocation are entangled after LSP even at high temperature. This reveals that precipitates are more likely to nucleate around the dislocations during dynamic aging in LSP. When the high-density dislocations and precipitates are entangled together, the movement of dislocation will be strongly pinned by precipitates, which would hinder dislocation slip and plastic deformation. In this way, the material strength increases, and the crack propagation rates slow down during fatigue test, which will ultimately increase the fatigue life.

Compared with the untreated, the sample after laser treatment would have a better fatigue performance due to the existences of the higher surface strength, deeper compressive residual stress, and higher stability of dislocation arrangement. However, the fatigue properties are more sensitive to defects in the material at high temperature. The fatigue life will be relatively reduced at high temperature compared with that at the ambient temperature. In addition, compressive residual stress also affects early fatigue crack path [50, 51]. The fatigue crack nucleation mechanism is that pore nucleates and grows between the inclusion and precipitated phase. Therefore, the cracking sources should propagate along the radial direction, but some of them axially extend. The result illustrates that compressive residual stress still affects early fatigue crack path during the residual stress relaxation, although the residual stresses will gradually release at 400 and 600 °C.

3.3.4 Conclusions

The stability of the residual stress, surface fatigue microstructure of LSP, and untreated ASTM: 410L 00Cr12 steel were investigated. LSP significantly enhances the isothermal stress-controlled fatigue behavior of ASTM: 410L 00Cr12 in the temperature range of 25–600 °C.

1. A deep layer of compressive residual stresses up to 1 mm with a maximum 400 MPa is introduced in the subsurface after LSP. The residual stress state and work hardening induced by LSP are more stable at elevated temperatures as compared to the untreated.
2. High-density dislocation and surface grain refinement play a leading role in increasing the high-temperature mechanical properties, and this is not affected by the ambient condition; LSP enhances fatigue life about 53.8 % in the low cycle regime at 10^5 cycles at elevated temperature up to 600 °C for the investigated material.
3. The fatigue lifetime of material after LSP is higher than that of untreated due to the existents of more stable near-surface microstructures, deformation-induced martensite, and the higher dislocation densities. Moreover, the pinning effect, which means surface grain refinement induced by LSP, is the main reason of prolonging the fatigue life at high temperatures.

LSP might be utilized in many age-hardening materials for improving the fatigue resistance at high temperatures. Therefore, it is necessary to compare the residual stress stability and fatigue behavior of the ASTM: 410L 00Cr12 steel to other different surface treatment technologies in the next studies.

3.4 Metallographic Structure Evolution and Dislocation Polymorphism Transformation of 6061-T651 Aluminum Alloy Processed by Laser Shock Peening: Effect of Tempering at the Elevated Temperatures

Abstract The aim of the study was to investigate the influences of laser shock processing (LSP) on metallographic structure evolution and dislocation configuration of 6061-T651 aluminum alloy at elevated temperature. Residual stress was determined by XRD by $\sin^2\psi$ method, optical microscopy (OM), and TEM observations of the grain, and precipitated phase evolutionary process after LSP was carried out. Surface topography and surface roughness were tested by a Surfcom 130A-Monochrome surface rough-meter. Morphologies of precipitated phases were monitored by SEM, and the dislocation configurations of samples after LSP were characterized by TEM. The results showed that the main mechanism of stress relaxation between 200 and 300 °C was the dislocation slip and dislocation climb, and dynamic recovery was the main mechanism during the stress relaxation deformation between 400 and 500 °C. The evolution mechanism of grain and precipitated phase of the 6061-T651 aluminum alloy between 25 and 500 °C was proposed. LSP was the principal element for inducing the "metallographic variation" (MV) effect of 6061-T651 aluminum alloy at the elevated temperature. The main strengthening mechanism of micro-hardness was dislocation strengthening and fine grain strengthening, and precipitated phase strengthening was the main strengthening mechanism at elevated temperature. "Dislocation polymorphism transformation" (DPT) effect was affirmed at elevated temperature, and the elevated temperature was the principal element for inducing the DPT effect of 6061-T651 aluminum alloy by LSP.

3.4.1 Introduction

LSP is a new surface modification technology, which uses high-strength impact wave of the interaction process between high-power density (GW/cm^2) lasers and materials to generate residual compressive stress and the high-density dislocation, and improves the mechanical properties of metal effectively [52], especially the metal materials of hardness [53], residual stress distribution [54], and resist fatigue life [55]. When the pressure of high-strength impact shockwave exceeds the dynamic yield strength of the metal, the shock wave can induce severe plastic deformation in the surface layer of the metals and alloys. Since lasers can be easily directed to fatigue-critical areas, LSP is expected to be widely applicable for improving the fatigue properties of alloys [56]. LSP also has the obvious advantages including no contact and heat-affected zone, excellent controllability, and impact [57]. 6061 aluminum alloys are widely used in automotive and aerospace applications because of their high properties, such as good strength [58], formability [59], and weldability and corrosion resistance [60–63]. The mechanical properties of aluminum alloy are often related to material internal microstructure [58–61], which are influenced by operating temperature.

Many researchers have investigated the effect on mechanical properties and microstructure of aluminum alloy, such as shot peening [64–66], cold rolling [67], LSP [68, 69], and oil jet peening [70]. Carvalho and Voorwald [66] investigated the effects of shot peening on fatigue strength of the 7050-T7451 Al alloy. Arun Prakash [70] evaluated the oil jet peening influence on the microstructure evolution and mechanical properties of aluminum alloy. Fribourg et al. [71] showed the effect of laser surface treatment on the microstructure of AA7449 aluminum alloy. Those above reports have just studied the effects of the mechanical properties on metal material at room temperature. Ren [72] reported the effects of LSP on heat-resistant steel of mechanical properties at high temperature; however, the influences of LSP on mechanical properties 6061-T651 aluminum alloy at elevated temperature have never been reported in the literature. The objective of this section is to investigate the effects of LSP on residual stress relaxation and the micro-mechanism, microstructure and precipitation behavior, surface topography, and micro-hardness of 6061-T651 aluminum alloy at the elevated temperature, and to provide a kind of reference method to understand and control the usability of aluminum alloy preferably. Furthermore, the strengthening mechanisms of LSP on micro-hardness and dislocation configuration evolution at elevated temperature are also revealed.

3.4.2 Experimental Procedures

3.4.2.1 Material and Technical Parameters

The 6061-T651 aluminum alloy is produced by a series of operations as follows: casting, homogenizing, and annealing. The specimens manufactured by 6061-T651 aluminum alloy were cut into a circular shape with dimensions of $\Phi16$ mm × 5 mm (diameter thickness). The chemical composition and photograph of 6061-T651 aluminum alloy are shown in Table 3.4 and Fig. 3.23a, respectively.

During LSP, the shock waves were induced by Nd:glass laser with a wavelength of 1064 nm and a pulse of 10 ns. The laser spot diameter was 3 mm, and the overlapping rate of the laser spot was 50 %. The water, with a thickness of about 2 mm, was used as the transparent confining layer, and the aluminum foil with a thickness of 0.12 mm was used as an absorbing layer to protect the specimen surface from thermal effects. The laser energy was 6 J. The parameters used in LSP are shown in Table 3.5 in detail. The specimen after LSP is shown in Fig. 3.23b.

Table 3.4 Chemical compositions of 6061-T651 aluminum alloy (wt%)

Cu	Mn	Mg	Zn	Cr	Ti	Si	Fe	Al
0.15–0.4	0.15	0.8–1.2	0.25	0.04–0.35	0.15	0.4–0.8	0.7	Balance

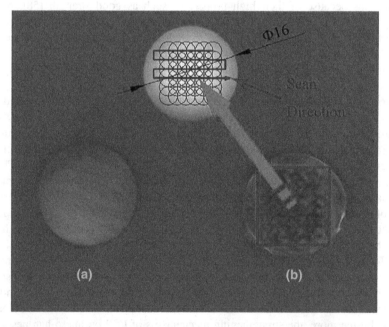

Fig. 3.23 Photograph of the specimens: **a** without LSP and **b** with LSP. Reprint from Ref. [73], Copyright 2013, with permission from Elsevier

Table 3.5 LSP parameters

Type	Value
Beam divergence of output/mrad	≤0.5
Pulse energy/J	6
Spot diameter/mm	3
Laser pulse width/ns	10
Repetition rate/Hz	5
Laser wavelength/nm	1064

And the samples were heated at the temperatures of 200, 300, 400, and 500 °C in chamber-type electric resistance furnace for 30 min, respectively, and then air cooled for 1 h.

3.4.2.2 Surface Topography and Roughness

Two-dimensional surface topography and surface roughness were tested by Surf-com 130A-Monochrome surface rough-meter. The arithmetic average deviation of profile (Ra) is treated as the main parameter to evaluate surface roughness. $R_a = (1/l)\int_0^1 y(x)dx$ [74], where l is the sample length and $y(x)$ is the deflection distance of profile. The tested length l was a 6 mm line segment along the laser spot center.

3.4.2.3 Measurements of Residual Stress

Residual stresses on the surface of specimens and through the thickness direction after LSP were determined by using XRD with $\sin^2\psi$ method. X-ray beam diameter was about 1 mm, X-ray source was CrKα ray, and the diffraction plane was a phase (311) plane. During the stress calculation, the Poisson's ratio was set to be 0.33. The feed angle of the ladder scanning was 0.1°/s. The scanning started angle and terminated angle were 142° and 135°, respectively. For the measurement of the residual stress along the depth direction, the electro-polishing material removal method was used. The measurement error was maintained with ±20 MPa, and as the error was more than 25 MPa measured again.

3.4.2.4 Measurements of Micro-Hardness

The micro-hardness was measured by HXD-1000TMSC/LCD Micro-hardness Tester. 0.49N was applied for micro-hardness measurements, and the loading time was 10 s. Micro-hardness results were obtained by taking the average values of ten different hardness measurements from each sample.

3.4.2.5 Microstructure Observations

The metallography structure of the samples by LSP were observed with OM after being etched by using Keller's reagent that consists of 2 ml of HF, 5 ml of HNO_3, 3 ml of HCl, and 190 ml of H_2O for 12 s at room temperature. Microstructural evolutions of the treated surface of samples after LSP at different temperatures were characterized by transmission electron microscopy using a JEM-2100(HR) transmission electron microscope operated at a voltage of 200 kV. The fabrication process of TEM foils was prepared by the following steps:

(i)   The treated samples were manufactured into thin foils using the without strain method;

(ii)   Thin foils were ground carefully to about 30 μm thickness;

(iii) Thin foils were thinned by the twin-jet electro-polishing.

3.4.3 Results and Discussions

3.4.3.1 Effects of LSP on Residual Stresses at Different Temperatures

Surface residual stress distributions of 6061-T651 aluminum alloy at different temperatures are shown in Fig. 3.24, which shows that the surface residual stress before and after LSP are −76 and −238 MPa, respectively.

With the increase of temperature, the surface residual stress decreases gradually; therefore, it could be found that the surface residual stress will be released at elevated temperature (up to 500 °C). The increase in amplitude of stress relaxation is large, and the residual stress reduces significantly in the temperature range of 25–300 °C. The residual compressive stress is −65 MPa at the temperature of 300 °C,

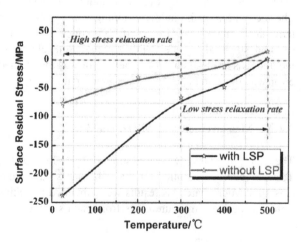

Fig. 3.24 Surface residual stress distributions at different temperatures. Reprint from Ref. [73], Copyright 2013, with permission from Elsevier

Fig. 3.25 Residual stress
distribution of the specimen
with distance. Reprint from
Ref. [73], Copyright 2013,
with permission from Elsevier

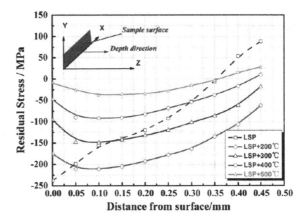

which has decreased about 57.6 % as compared with 25 °C. The increase in
amplitude tends to slow down when the temperature is above 300 °C. And residual
stress is 3 MPa at the temperature of 500 °C, which is the tensile stress. The residual
stress relaxation of the 6061-T651 aluminum alloy by LSP could be divided into
two phases (as shown in Fig. 3.25):

(i) the relaxation rate of residual stress is fast between 25 and 300 °C;

(ii) the relaxation rate tends to slow down above 300 °C, and the value of residual
     stress drops slowly. Figure 3.25 also shows the depth profiles of the residual
     stresses of the 6061-T651 aluminum alloy by LSP at different temperatures.
     The experimental results show that the higher the temperature, the faster the
     stress relaxation is. The stress relaxation is small as temperature is lowered to
     300 °C, but the stress relaxation becomes very large after 300 °C. The stress is
     almost relaxed completely at the temperature of 500 °C. It could be concluded
     that the affected layer of the 6061-T651 aluminum alloy by LSP has a higher
     magnitude of residual compressive stress from 25 to 300 °C. Therefore, LSP
     still has a beneficial effect on residual stress at the elevated temperature
     300 °C.

Thermal residual stress relaxation could be described by the Zener–Wert–Avrami function [53, 71]. According to the Zener–Wert–Avrami function, the equation
of temperature and residual stress could be inferred as follows:

$$\ln\left[\ln\left(\sigma_0^{RS}/\sigma^{RS}\right)\right] = m\ln(Bt_a) + m\frac{-\Delta H}{kT_a}$$

where

σ^{RS} is the residual stress at the surface after t_a hours of aging;

σ_0^{RS} is the initial value of the residual stress;

m is a parameter that depends on the dominating relaxation mechanism (m_{FWHM}
        as the reference value in this section);

Table 3.6 Coefficients of the Zener–Wert–Avrami of the residual stress relaxation behavior of LSP during elevated temperature

ΔH, eV	B, min^{-1}	k, eV/K	m_{FWHM}
1.35	3.06×10^{12}	8.617343×10^{-5}	0.22

t_a is the aging time;

A is a function of the material and the temperature shown in Table 3.6;

B is a constant;

k is the Boltzmann constant;

T_a is the aging temperature;

ΔH is the activation enthalpy for the relaxation process

From Eq. (3.1), the diagram of $\ln\left[\ln\left(\sigma_0^{RS}/\sigma^{RS}\right)\right]$, as a function of T_a for $t_a = 30$ min, is plotted in Fig. 3.26 according to the data from Table 3.6 [75, 76].

It is noted that the value of $\ln\left[\ln\left(\sigma_0^{RS}/\sigma^{RS}\right)\right]$ changes quickly between 25 and 300 °C, as shown in Fig. 3.26, while the value of $\ln\left[\ln\left(\sigma_0^{RS}/\sigma^{RS}\right)\right]$ after 300 °C does not have the strong change, which illustrates that the residual stress is released soon firstly and then tends to slow down. This releasing process is also in accord with the experimental results. Therefore, the effects of temperature on thermal residual stress relaxation could be described and analyzed precisely using the Zener–Wert–Avrami function. Although the residual compressive stress will be relaxed at elevated temperature, LSP changes the residual stress at the specimen surface, and it is obviously higher than the surface residual stress of the specimen without LSP. High amplitude of the residual stress would resist against the fatigue crack initiation and growth [53], which is important to improve the fatigue property [77] and fatigue life [78] of the component.

Fig. 3.26 Influence of temperature on residual stress in $\ln\left[\ln\left(\sigma_0^{RS}/\sigma^{RS}\right)\right] - T_a$ diagram. Reprint from Ref. [73], Copyright 2013, with permission from Elsevier

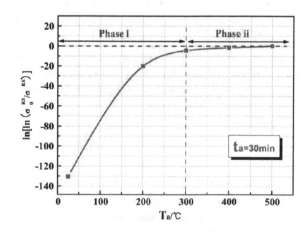

3.4.3.2 Micro-Mechanism Changing of Residual Stress Relaxation

Figure 3.27 shows the TEM micrographs of the structure of the 6061-T651 aluminum alloy before LSP and after LSP. It could be observed from Fig. 3.27a that the precipitated phases are distributed in the trans-granular or along the grain boundary, and there are no obvious dislocations and dislocation lines. Figure 3.27b is the magnified image of the circle in Fig. 3.27a, showing the morphology of precipitated phases. The main morphology of precipitated phase is granular and baculiform. Figure 3.27c, d shows the TEM observations of the structure after LSP, in which four typical deformation-induced microstructure features are identified, including dislocation line, dislocation walls, dislocation network, and dislocation tangle.

At elevated temperature, because the stacking fault energy of aluminum alloy will increase and the self-diffusion activation energy will decrease, the dislocation slip and dislocation climb are easy to occur. When the polygonal sub-grain structure is formed during the deformation process, namely dynamic recovery is occurring [79]. Essentially, dynamic recovery is the result of the dislocation slip, the dislocation climb, and the reduction of dislocation density [80]. The TEM graphs after residual stress relaxation of 6061-T651 aluminum alloy at different temperatures are

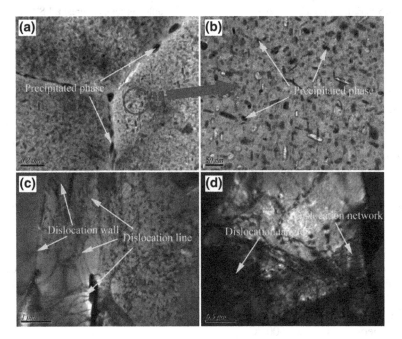

Fig. 3.27 TEM micrographs of 6061-T651 aluminum alloy before LSP and after LSP. **a** Microstructure of 6061-T651 aluminum alloy before LSP; **b** the magnifications of the *circle* in (**a**); **c** and **d** microstructure of 6061-T651 aluminum alloy after LSP. Reprint from Ref. [73], Copyright 2013, with permission from Elsevier

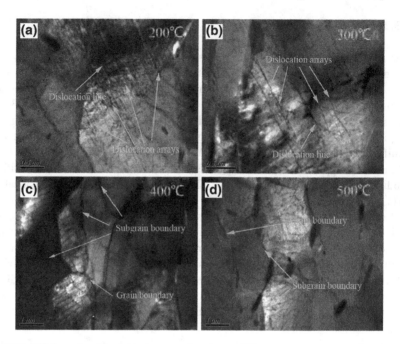

Fig. 3.28 TEM graphs after residual stress relaxation at different temperatures. **a** LSP +200 °C; **b** LSP +300 °C; **c** LSP +400 °C; and **d** LSP +500 °C. Reprint from Ref. [22], Copyright 2013, with permission from Elsevier

shown in Fig. 3.28. Figure 3.28a, b shows that there are plenty of dislocations, which are long, straight, and parallel dislocations. At the same time, the parallel dislocations have developed the dislocation array, and the dislocation density is high. It cannot be observed that the dislocation piles up in the grain boundary, which indicates that the dislocation climbing takes place in the grain boundary. Hence, it could be concluded that the main mechanism of stress relaxation from 200 to 300 °C is the dislocation slip and dislocation climb. The density of dislocation slip lines at 200 °C is lower than that at 300 °C, indicating that the later micro-plastic deformation is bigger than the former. That is to say, the stress relaxation is more intense than that at 300 °C. Figure 3.28c, d shows that an obvious recovery process has occurred in the relaxed specimens and the dislocation density is lower than that shown in Fig. 3.28a, b. The quantity of dislocations appears to be decreased in comparison with the unrelaxed specimens after LSP (Fig. 3.27c, d). The dislocation motion is blocked by either grain boundary or precipitated phase. Moreover, the sub-grain boundaries are also observed clearly from Fig. 3.28c, d. So, dynamic recovery is the main mechanism during the stress relaxation deformation between 400 and 500 °C.

The result of Fig. 3.24 shows that the stress relaxation rate is large when the temperature is lowered to 300 °C, while the stress relaxation rate tends to slow down after 300 °C. When the temperature is lowered to 300 °C, plenty of mobile dislocations change to the slippage and climb during the process of adjusting internal stress. In the meantime, the dislocation forms the parallel arrangement to offset their own stress field effect, which shows that the stress relaxation rate is high on the microscopic scale. The mobile dislocations decrease, and the dislocation density reduces as the temperature increases. The dislocation tangles have been adjusted when the temperature surpasses to 400 °C. In the meantime, increasing temperature (up to 500 °C) results in an impressive increase of dislocation motion resistance and the dislocation configurations tend to be steady, and then develop the clear sub-grain boundary. This result illustrates that stress relaxation rate tends to slow down in macro-level, namely the residual stress is approaching to the stress relaxation limit. From these results, it could be deduced that LSP, which produces plentiful mobile dislocations, plays an important role in the micro-mechanism changing of 6061-T651 aluminum alloy at elevated temperature.

3.4.3.3 Effects of LSP on Grain and Precipitated Phase

Figure 3.29 exhibits the OM morphologies of 6061-T651 aluminum alloy treated at different temperatures. It could be observed from Fig. 3.29a that the original structure of the 6061-T651 aluminum alloy without LSP is thick, and distributes unequally. The precipitated phases shown in Fig. 3.29a are sparsely distributed in the trans-granular or along the grain boundary, which are also shown in Fig. 3.27a. The original structure after LSP is obviously refined, and the grain size decreases and distributes equably, as shown in Fig. 3.29b. The different temperatures lead to the recrystallization of original structure in some degree, as shown in Fig. 3.29c–e, respectively. It could be observed from Fig. 3.29c that the size of primitive grains and precipitated phases begin to decrease. And a handful of recrystallized grains are found in the grain boundary at 200 °C. Figure 3.29d shows, at the temperature of 300 °C, the grain size is getting smaller, the number of dynamic recrystallization is growing, and most of the precipitated phases have been integrated into the aluminum alloy substrate. Few small precipitated phases are observed in the transgranular or along the grain boundary. The grain size begins to become larger, and the number of precipitated phases begins to grow at the temperature of 400 °C, which are shown in Fig. 3.29e. The primitive grain is phagocytized by the new grain, and the dynamic recrystallization is completed at 500 °C, as shown in Fig. 3.29f.

The grain size is essentially identical and much bigger than before. In the meantime, the precipitated-phase particles are thick and distribute in the transgranular or along the grain boundary equably. The size of grain and precipitated phase decreases originally and amplifies as the temperature increases, while the quantity of precipitated phase increases firstly, and then reduces. This phenomenon could be due to the fact that LSP changes the microstructure of material surface, which leads the grain and precipitated phase to grow up abnormally.

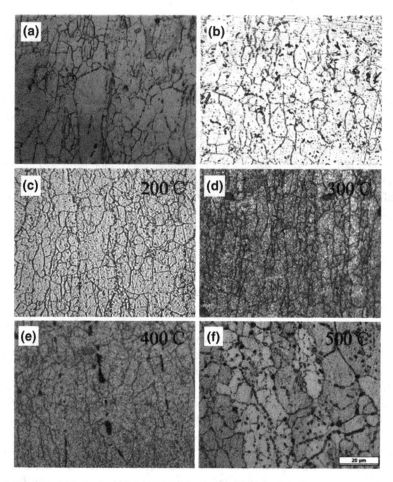

Fig. 3.29 OM morphologies of 6061-T651 aluminum alloy at different temperatures **a** without LSP; **b** LSP; **c** LSP +200 °C; **d** LSP +300 °C; **e** LSP +400 °C; and **f** LSP +500 °C. Reprint from Ref. [73], Copyright 2013, with permission from Elsevier

3.4.3.4 Metallographic Structure Evolution at Elevated Temperature

LSP induces a mass of strain effects near material surface, such as dislocation lines, dislocation network, and dislocation tangles (as shown in Fig. 3.27c and d). LSP also produces the plentiful mobile dislocations and sessile dislocations (pinned). In high temperature condition, the size of grain and precipitated phase are not with the increasing of temperature growing or lessening normally because of the dislocation motion, recombination and the pinning effect of precipitated phase, this phenomenon was named as the "metallographic variation" (MV) effect induced by the laser shock processing at the elevated temperature. The schematic diagram of grain and precipitated phase evolutionary process of 6061-T651 aluminum alloy is shown in

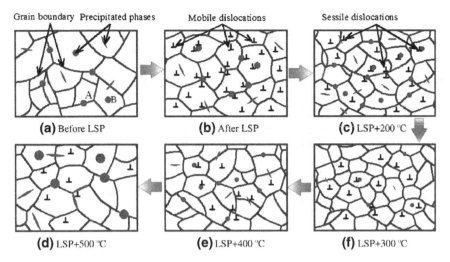

Fig. 3.30 Schematic of grain and precipitated phase evolutionary process at different temperatures. Reprint from Ref. [73], Copyright 2013, with permission from Elsevier

Fig. 3.30. LSP induces a lot of mobile dislocations and sessile dislocations at the 6061-T651 aluminum alloy surface, as shown in Fig. 3.30b and c. It is well known that different dislocation motions cause the primitive grain of the surface layer to form the dislocation walls and dislocation tangles, as shown in Fig. 3.27c, d. And these dislocation walls separate the thick grains into the different sub-grains, which leads to the thick grains to be refined. The precipitated phases distributing in the trans-granular or along the grain boundary (as shown in the A and B in Fig. 3.30a) have a strong pinning effect. In this case, the dislocation motion and crystal boundary migration are blocked by precipitated phases seriously. The system is in the metastable state [81], as the precipitated-phase size is equal to the average size. Therefore, there is a solute concentration gradient which is from high to low between the small particles and big particles. The solute around small particles has a tendency to diffuse to near the big particles. Figure 3.30c, d shows that, with the increasing temperature between 200 and 300 °C, the elevated temperature accelerates the diffusion process and the metastable state of system is destroyed. The solute concentration around small particles is lower than the solute concentration which is in a metastable state, as the influences of temperature. Hence, precipitated phase happens to dissolve and diffuse around the big precipitated-phase particles. It could be found that the size and the number of precipitated phase start to lessen. The pinning effect is going to be weakening due to the decrease of precipitated-phase size. Therefore, the sessile dislocations are going to release, and then change into mobile dislocations. The mobile dislocations get through the migration and restructuring to form dislocation walls, and then develop the sub-grain. Consequently, the average size decreases. The total number of dislocations is fewer and fewer for the mobile dislocations developing the sub-grain; this could be clearly observed from Fig. 3.30c, d. When the temperature is up to 400 °C, the solute atoms diffuse to around big particles, and the big particles grow up gradually as the solute concentration is higher than the concentration of big particles. At the same time, the increase of nucleation activation energy promotes the precipitated phases to dissolve out. The elevated temperature (exceeding 400 °C) increases the thermal activation energy of 6061-T651 aluminum alloy and leads to the crystal boundary migration rate to be accelerated, and the grain boundary breaks away from the thermodynamic free energy produced by precipitated phase. In addition, the distortional strain energy induced by LSP releases at high temperature, which would promote the grains to grow up until the moving grain boundary, is pinned by the precipitated phase again, as shown in Fig. 3.30e, f. In terms of the above experimental observations and theoretical analyses, it could be concluded that LSP is the principal element for inducing the "metallographic variation" (MV) effect of 6061-T651 aluminum alloy at the elevated temperature.

3.4.3.5 Surface Topography and Roughness

The surface morphology of the metal material has a great effect on fatigue behavior [82]. Full investigation of surface topography by LSP treatment at elevated temperature has never been conducted in the previous studies. It is crucial to study the effect of LSP on the surface topography of 6061-T651 aluminum alloy. Figure 3.31 shows the 2D surface topography of 6061-T651 aluminum alloys with LSP and un-LSP.

The microscopic surface without LSP is very smooth, whose height of peak and valley scatters from −0.5 to 0.5 μm. The height of peak and valley is enlarged rapidly (from −2.5 to 2.5 μm) after LSP, which illustrates that LSP changes the surface topography. The 2D surface topography of the 6061-T651 aluminum alloy by LSP in the temperature range of 200–500 °C is shown in Fig. 3.32, which shows

Fig. 3.31 2D surface topography of 6061-T651 aluminum alloy. Reprint from Ref. [83], Copyright 2013, with permission from Elsevier

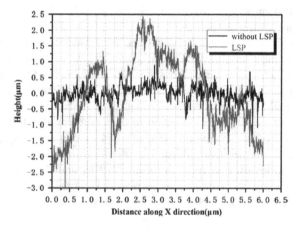

Fig. 3.32 2D surface
topography of 6061-T651
aluminum alloy at different
temperatures. Reprint from
Ref. [83], Copyright 2013,
with permission from Elsevier

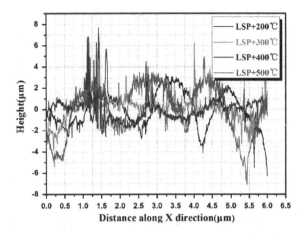

the absolute value of the microscopic surface height of peak and valley increases gradually with the temperature increase.

But all of the microscopic height distributes from −5 to 7 μm, illustrating that elevated temperature has a little change on the 2D surface topography of 6061-T651 aluminum alloy after LSP between 200 and 500 °C.

It could be observed from Fig. 3.33 that the surface roughness of 6061-T651 increases gradually in the temperature range of 25–500 °C. The surface roughness after LSP at 200, 300, 400, and 500 °C is 0.387, 0.414, 0.425, and 0.444 μm, respectively.

LSP surface roughness above 200 °C is higher than that at room temperature, while the surface roughness remains constant after 200 °C. The reason is that 6061-T651 aluminum alloy is oxidized at elevated temperature, which leads the sample surface to form an uneven oxidation layer and increase the surface roughness. Therefore, the surface roughness at elevated temperature is enlarged compared to that at room temperature.

Fig. 3.33 Surface roughness
with LSP and without LSP at
different temperatures.
Reprint from Ref. [83],
Copyright 2013, with
permission from Elsevier

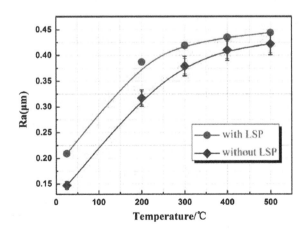

As the temperature is further increased in high-temperature environment (up to 500 °C), the height of peak and valley at the surface oxidation layer remains constant; that is to say, there are few changes of the surface roughness. Hence, the surface roughness of the specimen at room temperature and elevated temperature was not exceeding 0.21 and 0.45 μm, respectively. The smooth surface would lead to significant improvements in resistance to fatigue crack initiation [84].

3.4.3.6 Micro-Hardness Changing

The results of the surface micro-hardness at different temperatures are shown in Fig. 3.34.

Micro-harnesses by un-LSP and LSP are 160.6 and 205.4 HV, respectively, at room temperature. This result illustrates that LSP could increase micro-hardness obviously, which is also proved by He [85] and Xiong [86].

LSP also improves the micro-hardness of 6061-T651 aluminum alloy at elevated temperature compared with un-LSP. With the temperature increase, the micro-hardness decreases to 133.6 HV firstly, and then increases up to 215.5 HV. The least value and the peak value are observed at 300 and 500 °C, respectively. It could be concluded that LSP still has a beneficial effect on micro-hardness at elevated temperature (up to 500 °C).

Figure 3.35 shows the depth profiles of micro-hardness of 6061-T651 aluminum alloy at different temperatures.

The micro-hardness of sample by un-LSP is constant, while the micro-hardness by LSP decreases gradually with the increasing depth. The micro-hardness by LSP is higher than that with un-LSP from 0 to 0.35 mm. As the depth exceeds 0.35 mm, the micro-hardness would be equivalent again. Therefore, the results illustrate that the LSP-affecting depth is about 0.35 mm. This result is attributed to the high-power density and short-pulsed laser beam to irradiate metal surface and develop the huge compressive stress wave to impact metal material [87]. When the depth exceeds the effecting range, the micro-hardness of material would not change any longer.

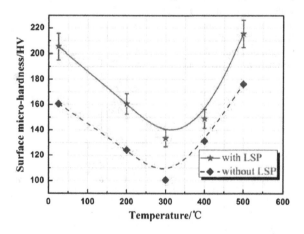

Fig. 3.34 Comparison of micro-hardness at different temperatures. Reprint from Ref. [83], Copyright 2013, with permission from Elsevier

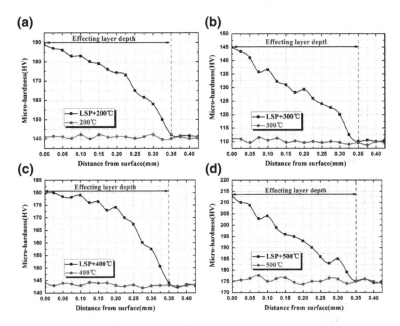

Fig. 3.35 Micro-hardness profile on the cross section. Reprint from Ref. [83], Copyright 2013, with permission from Elsevier

The plastically affected depth L during LSP could be calculated by Peyre et al. [88].

$$L = \frac{C_e C_p t}{C_e - C_p} \left(\frac{P - \text{HEL}}{2\text{HEL}} \right)$$

where C_e and C_p are the elastic and plastic shock wave velocities, respectively, $C_e = 6.39 \times 10^6$ mm/s, $C_p = 5.33 \times 10^6$ mm/s [76], HEL is the Hugoniot elastic limit, and t is the pressure pulse duration. The pressure P could be estimated by the following relationship [75]:

$$P = 0.01 \sqrt{\frac{\alpha}{2\alpha}} = \sqrt{z} \sqrt{I_0}$$

where α is the efficiency of the interaction, and a is 0.1. I_0 is the incident laser power density, and the value is 8.492 GW/cm², and Z is the reduced shock impedance between the target and water constraint layer, which could be defined as

$$\frac{2}{Z} = \frac{1}{Z_{\text{target}}} + \frac{1}{Z_{\text{water}}}$$

As to aluminum target and water constraint layer,

$$Z_{\text{target}} = 1.38 \times 10^6 \text{ gcm}^{-2}\text{s}^{-1}$$
$$Z_{\text{water}} = 0.165 \times 10^6 \text{ gcm}^{-2}\text{s}^{-1}$$

It could be deduced that the plastically affected depth L is about 0.41 mm, which is consistent with the test result. The theory result and experimental result show the reinforcement depth of the 6061-T651 aluminum alloy by LSP is 0.35 mm at elevated temperature. It could be found that the improvements of the surface micro-hardness distribution, which illustrates LSP still has a beneficial reinforcing effect, and LSP improve the fatigue property of material at elevated temperature, as shown in Figs. 3.34 and 3.35.

3.4.3.7 Strengthening Mechanism

Figure 3.36 shows the TEM of 6061-T651 aluminum alloy. As shown in Fig. 3.36a, a small number of precipitated phases were distributed in the trans-granular, and there was no obvious dislocation network. High-density dislocation line, dislocation network, and sub-grain boundary could be clearly observed in Fig. 3.36b. Figure 3.36b also indicates that the surface by LSP is generated plastic deformation, the dislocation density is obviously improved, and the grain is refined.

The yield strength $\Delta\sigma_y$ of the deformed metal is connected with the dislocation density ρ of dislocations, according to the Bailey–Hirsch function $\Delta\sigma_y = \sigma_0 + \alpha_2 Gb\rho^{1/2}$ (where $\Delta\sigma_y$ is the yield strength, σ_0 is the yield strength of the undeformed metal, α_2 is a material constant, G is the shear modulus, and b is the length of the Burgers vector) [89]; the higher the dislocation density ρ, the bigger the yield strength $\Delta\sigma_y$, namely the micro-hardness would be increased in macro-level. At the same time, it could be deduced from the Hall–Petch function $H_v = H_{v0} + K_{Hv}d^{-1/2}$ (where H_{v0} is the intrinsic hardness, K_{Hv} is the Hall–Petch coefficient, and d is the

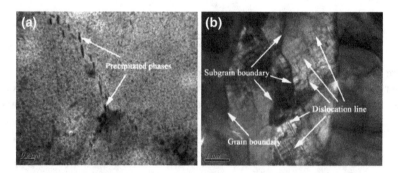

Fig. 3.36 TEM of 6061-T651 aluminum alloy **a** before LSP and **b** after LSP. Reprint from Ref. [83], Copyright 2013, with permission from Elsevier

average grain size) [90] that the smaller the grain size, the bigger the micro-hardness, within the current grain size range. Therefore, the main strengthening mechanism of LSP on micro-hardness was dislocation strengthening and fine grain strengthening at the room temperature.

Surface hardness is closely related to the morphology, size, and quantity of precipitated phase [91, 92]. However, the morphology, size, and quantity of precipitated phase also relate to the temperature [92]. The transformation kinetics is increased for the elevated temperature, and the critical nucleus size of precipitated phase is decreased as the nucleation rate improves. Meanwhile, the quantity of precipitated phase is increased. Figure 3.37 shows that the morphologies of precipitated phases in 6061-T651 aluminum alloy after LSP at different temperatures.

It could be found that the main morphology of precipitated phases is granular and baculiform. The dislocation strengthening effect would weaken gradually at elevated temperature, as shown in Fig. 3.38, due to the reduction of dislocation number and the decrease of dislocation density. It could be clearly observed from Fig. 3.37 that the size of precipitated phase decreases originally, and then increases with the temperature increase. The bigger the precipitated-phase size, and the smaller the separation distance, the better the strengthening effect. Meanwhile, it could be deduced that the main strengthening mechanism of LSP on the micro-hardness was precipitated phase strengthening at elevated temperature combining with the result of Fig. 3.34, which is the primary reason why the surface hardness decreases originally, and then increases.

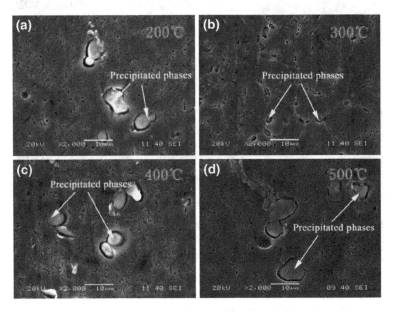

Fig. 3.37 Morphologies of precipitated phases in 6061-T651 aluminum alloy after LSP at different temperatures: **a** LSP +200 °C; **b** LSP +300 °C; **c** LSP +400 °C; and **d** LSP +500 °C. Reprint from Ref. [32], Copyright 2013, with permission from Elsevier

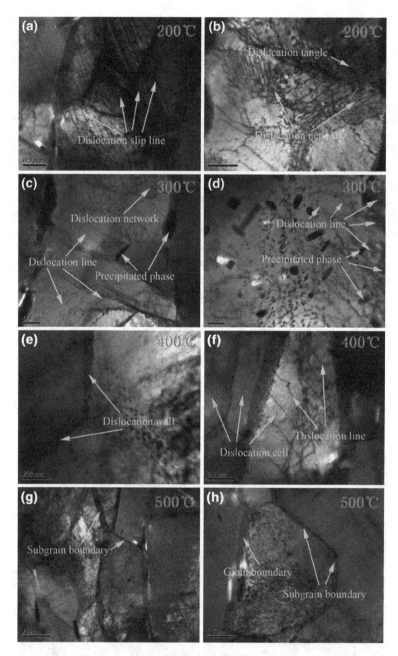

Fig. 3.38 Typical TEM images of 6061-T651 aluminum alloy subjected to LSP at different temperatures: **a**, **b** LSP +200 °C; **c**, **d** LSP +300 °C; **e**, **f** LSP +400 °C; **g**, **h** LSP +500 °C. Reprint from Ref. [32], Copyright 2013, with permission from Elsevier

3.4.3.8 Dislocation Configuration

LSP induces high density of dislocation structures, such as dislocation line and dislocation tangle. High density of dislocation would change upper state into a lower state at elevated temperature, and the adjusting dislocation changes into the different dislocation configurations orderly (dislocation tangle, dislocation network, dislocation wall, and dislocation cell) with the increasing temperature, and finally develops the clear sub-grain boundary, and we named this phenomenon as the DPT effect induced by the elevated temperature. During this experiment, DPT effect was found at elevated temperature.

Figure 3.38 shows the typical TEM images of 6061-T651 aluminum alloy subjected to LSP impact at different temperatures. It could be observed clearly from Fig. 3.38a that there are plenty of dislocations and dislocation slip lines at the temperature of 200 °C. The dislocation slip lines are long, straight, and parallel. And the dislocation slip lines have developed the dislocation array at the same temperature. Therefore, the main deformation mechanism is the dislocation slip at the temperature of 200 °C.

As the grain is subjected to the external stress induced by LSP, a large number of dislocations are produced because of transformation. The dislocations develop the dislocation tangle or the dislocation network, as shown in Fig. 3.38b. A large number of dislocations change from upper state into lower state at 300 °C, which cause the dislocation tangle to expand and discretize, as shown in Fig. 3.38c.

The precipitated phases (shown as Fig. 3.38d) distributing in the trans-granular and along the grain boundary have a strong pinning effect, causing the increase of dislocation density near precipitated phases. In the meantime, precipitated phases block the dislocation motion and the crystal boundary migration seriously. And the recrystallization process is delayed as a result. With the increasing of temperature, the dislocation was arranged regularly at 400 °C. The dislocation attracts each other and then disappears, and the dislocation with different nature arranges regular, but then develops dislocation wall, as shown in Fig. 3.32e.

Figure 3.38f shows that the dislocation configuration exhibits in the shape of dislocation cell and the tangled dislocation gradually develop into the cell structure with boundary. It could be observed from Fig. 3.38g that the elevated temperature improves the thermal activation energy and causes the accelerating crystal boundary migration rate.

The elevated temperature provides more energy for dislocation movement, and the elevated temperature accelerates the speed of the dislocation movement. The dislocation in grain moves to the cell wall continuously. As a result, dislocation density reduces greatly; dislocation configuration tends to be stable and forms the clear grain boundary. The high-density dislocation in grain is shown in Fig. 3.38h. The reason is considered that the transformation occurs during the recrystallization process, and the high-density dislocation existing in grain improves the distortion strain energy, which promotes the dynamic recrystallization. It could be concluded that the elevated temperature was the principal element for inducing the DPT effect of 6061-T651 aluminum alloy by LSP.

3.4.4 Conclusions

The influences of laser shock processing on residual stress relaxation and metallographic structure evolution of 6061-T651 aluminum alloy at elevated temperature are investigated in this section. Some important conclusions could be made as follows:

1. The affected layer of the 6061-T651 aluminum alloy by LSP has a higher magnitude of residual compressive stress between 25 and 300 °C, and LSP still has a beneficial effect on residual stress at the elevated temperature of 300 °C. The effects of temperature on thermal residual stress relaxation could be described and analyzed precisely by the Zener–Wert–Avrami function.
2. The main mechanism of stress relaxation between 200 and 300 °C is the dislocation slip and dislocation climb, and dynamic recovery is the main relaxation mechanism during the stress relaxation deformation between 400 and 500 °C.
3. LSP is the principal element for inducing the "metallographic variation" (MV) effect of 6061-T651 aluminum alloy at the elevated temperature. The size of grain and precipitated phase decreases originally with the temperature increasing.
4. LSP changed the surface topography of 6061-T651 aluminum alloy, but all of the microscopic height distributed from −5 to 7 μm after LSP between 200 and 500 °C.
5. The affected layer of the 6061-T651 aluminum alloy by LSP has a great improvement at the elevated temperature. The main strengthening mechanism of LSP on the micro-hardness was dislocation strengthening. Fine grain strengthening and precipitated phase strengthening were the main strengthening mechanism at elevated temperature.
6. "Dislocation polymorphism transformation" (DPT) effect was found at elevated temperature. The elevated temperature was the principal element for inducing the DPT effect of 6061-T651 aluminum alloy by LSP. The results provide a kind of reference method to understand and control the usability of aluminum alloy preferably.

References

1. Hong SG et al (2004) Dynamic strain aging under tensile and LCF loading conditions, and their comparison in cold worked 316L stainless steel. J Nucl Mater 328:232–242
2. Hong SG et al (2007) Temperature effect on the low-cycle fatigue behavior of type 316L stainless steel: Cyclic non-stabilization and an invariable fatigue parameter. Mater Sci Eng A 457:139–147
3. Vogt JB et al (2002) High temperature low cycle fatigue of 2.25Cr1Mo steels: role of microstructure and effect of environment. Solid Mech Mater Eng 45:46–50
4. Wu LL, Holloway BC (2000) Analysis of diamond-like carbon and Ti/MoS2 coatings on Ti6Al4V substrates for applicability to turbine engine applications. Surf Coat Technol 130:207–213

5. Suh CM et al (1990) Fatigue microcracks in type 304 stainless steel at elevated temperature. Fatigue Fract Eng Mater Struct 13:487–496
6. Fan ZC et al (2007) Fatigue-creep behavior of 1.25Cr0.5Mo steel at high temperature and its life prediction. Int J Fatigue 29:1174–1183
7. Bethge K et al (1990) Crack initiation and crack propagation under thermal cyclic loading. High Temp Technol 8:98–104
8. Ren WJ et al (2005) Evaluation of coatings on Ti6Al4V substrate under fretting fatigue. Surf Coat Technol 192:177–188
9. Hutson AL et al (2002) Effect of various surface conditions on fretting fatigue behavior of Ti6Al4V. Int J Fatigue 24:1223–1234
10. Montros CS et al (2002) Laser shock processing and its effects on microstructure and properties of metal alloys: a review. Int J Fatigue 24:1021–1036
11. Romain JP et al (1986) Laser shock experiments at pressures above 100Mbar. Physica 139:595–598
12. Hyukjae L, Shankar, M (2004) Stress relaxation behavior of shot peening Ti-6Al-4V under fretting fatigue at elevated temperature. Mater Sci Eng A 366:412–420
13. Hyukjae L et al (2005) Investigation into effects of re-shot-peening on fretting fatigue behavior of Ti6Al4V. Mater Sci Eng A 390:227–232
14. Venkatech V, Rack HJ (1999) A neural network approach to elevated temperature creep-fatigue life prediction. Int J Fatigue 21:225–234
15. Shang DG et al (2007) Creep-fatigue life prediction under fully-reversed multiaxial loading at high temperatures. Int J Fatigue 29:705–712
16. Ren XD et al (2010) Effects of laser shock processing on 00Cr12 mechanical properties in the temperature range from 25 to 600 °C. Appl Surf Sci 257:1712–1715
17. Ling P, Wight, CA (1995) Laser-generated shock waves in thin films of energetic materials. J Appl Phys 78:7022–7025
18. Couturier S et al (1996) Shock profile induced by short laser pulses. J Appl Phys 79:9338–9342
19. Montross CS et al (2000) Subsurface properties of laser peened 6061-T6 Al weldments. Surf Eng 16:116–121
20. Ren XD et al (2011) Mechanical properties and residual stresses changing on 00Cr12 alloy by nanoseconds laser shock processing at high temperatures. Mat Sci Eng A 528:1949–1953
21. SAE AMS 2546 (2004) http://www.sae.org. Accessed 9 Nov 2012
22. Montross CS et al (2002) Laser shock processing and its effects on microstructure and properties of metal alloys: a review. Int J Fatigue 24:1021–1036
23. Zhang YK et al (2001) Mechanism ofimprovement on fatigue life of metal by laser-excited shock waves. Appl Phys A 72:113–116
24. Zhang YK et al (2009) Effect of laser shockprocessing on the mechanical properties and fatigue lives of the turbojet engine bladesmanufactured by LY2 aluminum alloy. Mater Des 30:1697–1703
25. Wang F et al (2007) Materials 14:529–532
26. Wang SP et al (1998) Compressive residual stressintroduced by shot peening. J Mater Process Technol 73:64–73
27. Farrahi GH et al (1995) Effect of shot peening on residual stressand fatigue life of spring steel. Fatigue Fract Eng Mater Struct 18:211–220
28. Guagliano M, Vergani L (2004) An approach for prediction of fatigue strength of shot peened components. Eng Fract Mech 71:501–512
29. Schulze V (2006) Modern mechanical surface treatment: states, stability, effects. Wiley-VCH, Weinheim
30. Juijerm P and Altenberger I (2007) Effect of temperature on cyclic deformation behavior and residual stress relaxation of deep rolled under-aged aluminium alloy AA6110. Mater Sci Eng A 452–453:475–482
31. Hatamleh O et al (2009) An investigation of the residual stress characterization and relaxation in peened friction stir welded aluminium lithium alloy joints. Mater Des 30:3367–3373

32. James MR (1985) Residual stress and stress relaxation. Plenum, New York
33. Altenberger I et al (2001) Analysis and assessment of residual stress states in mechanically surface treated materials. Mater Sci Res Int 1:275
34. Nalla RK et al (2003) On the influence of mechanical surface treatments-deep rolling and laser shock peening-on the fatigue behavior of Ti-6Al-4V at ambient and elevated temperatures. Mater Sci Eng A 355:216–230
35. Masaki K et al (2007) Effects of laser peening treatment on high cycle fatigue properties of degassing-processed cast aluminum alloy. Mater Sci Eng A 468–470:171–175
36. Nikitin I, Altenberger I (2007) Comparison of the fatigue behavior and residual stress stability of laser-shock peened and deep rolled austenitic stainless steel AISI304 in the temperature range 25–600 °C. Mater Sci Eng A 465:176–182
37. Heitkemper M et al (2003) Fatigue and fracture behavior of a laser surface heat treated martensitic high-nitrogen tool steel. Int J Fatigue 25:101–106
38. Ye BC et al (2010) Warm laser shock peening driven nanostructures and their effects on fatigue performance in aluminum alloy 6160. Adv Eng Mater 12:291–297
39. SAE AMS 2546 (2004) http://www.sae.org. Accessed 9 Nov 2012
40. Ren NF et al (2014) High temperature mechanical properties and surface fatigue behavior improving of steel alloy via laser shock peening. Mat Des 53:452–456
41. Scholtes B (1997) Assessment of residual stresses. In: Structural and residual stress analysis by nondestructive methods. Amsterdam, pp 590–632
42. Nikitin I et al (2004) High temperature fatigue behavior and residual stress stability of laser-shock peened and deep rolled austenitic steel AISI 304. Scripta Mater 50:1345–1350
43. Altenberger I et al (1999) Cyclic deformation and near surface microstructures of shot peened or deep rolled austenitic stainless steel AISI 304. Mater Sci Eng A 264:1–16
44. Xuan FZ et al (2010) Mass transport in laser surface nitriding involving the effect of high temperature gradient: simulation and experiment. Comput Mater Sci 49:104–111
45. Juijerm P, Altenberger I (2007) Effect of temperature on cyclic deformation behavior and residual stress relaxation of deep rolled under-aged aluminium alloy AA6110. Mater Sci Eng A 452–453:475–482
46. Juijerm P, Altenberger I (2006) Residual stress relaxation of deep-rolled Al-Mg-Si-Cu alloy during cyclic loading at elevated temperatures. Scripta Mater 55:1111–1114
47. Xuan FZ et al (2010) Laser surface nitriding of Ti6Al4V alloy coupled with an external stress field. J Mater Res 25:344–9
48. Konig GW (2002) In: Proceedings of the 8th international conference on shot peening (ICSP), pp 13–22
49. Meyers MA et al (2001) The onset of twinning in metals: a constitutive description. Acta Mater 49:4025–39
50. Ren XD et al (2013) The effects of residual stress on fatigue behavior and crack propagation from laser shock processing-worked hole. Mater Des 44:149–54
51. Ren XD et al (2011) Comparison of the simulation and experimental fatigue crack behaviors in the nanoseconds laser shocked aluminum alloy. Mater Des
52. Yang JM et al (2001) Laser shock peening on fatigue behavior of 2024-T3 Al alloy with fastener holes and stopholes. Mater Sci Eng A 298:296–299
53. Rubio- González C et al (2004) Effect of laser shock processing on fatigue crack growth and fracture toughness of 6061-T6 aluminum alloy. Mater Sci Eng A 386:291–295
54. Nikitin I et al (2004) High temperature fatigue behavior and residual stress stability of laser-shock peened and deep rolled austenitic steel AISI 304. Scr Mater 50:1345–1350
55. Schubbe JJ (2009) Fatiguecrackpropagationin7050-T7451platealloy. Mech Eng Fract 76:1037–1048
56. Rubio-González C et al (2011) Effect of laser shock processing on fatigue crack growth of duplex stainless steel. Mater Sci Eng A 528:914–919
57. Sudha C et al (2010) Mater Manuf Process 25(2010):956
58. Ahmad Fauzi MN et al (2010) Microstructure and mechanical properties of alumina-6061 aluminum alloy joined by friction welding. Ismail, Mater Des 31:670

59. Maisonnette D et al (2011) Effects of heat treatments on the microstructure and mechanical properties of a 6061aluminium alloy. Mater Sci Eng A 528:2718

60. Uday MB et al (2011) Effect of welding speed on mechanical strength of friction welded joint of YSZ–alumina composite and 6061 aluminum alloy. Mater Sci Eng A 528:4753

61. Fahimpour V et al (2012) Corrosion behavior of aluminum 6061 alloy joined by friction stir welding and gas tungsten arc welding methods. Mater Des 39:329

62. Wu KH et al (2007) Thermal stability and corrosion resistance of polysiloxane coatings on 2024-T3 and 6061-T6 aluminum alloy. Surf Coat Technol 201:5782

63. El-Menshawy K et al (2012) Corros Sci 54(2012):167

64. Carvalho ALM, Voorwald HJC (2007) Influence of shot peening and hard chromium electroplating on the fatigue strength of 7050-T7451 aluminum alloy. Int J Fatigue 29:1282

65. Gao YK, Wu XR (2011) Experimental investigation and fatigue life prediction for 7475-T7351 aluminum alloy with and without shot peening-induced residual stresses. Acta Mater 59:3737

66. Luong H, Hill MR (2010) The effects of laser peening and shot peening on high cycle fatigue in 7050-T7451 aluminum alloy. Mater Sci Eng A 527:699–707

67. Wang D et al (2009) Effect of recrystallization and heat treatment on strength and SCC of an Al-Zn-Mg-Cu alloy. Mater Chem Phys 117:228

68. Zhang YK et al (2001) Elastic properties modification in aluminum alloy induced by laser-shock processing. Mater Sci Eng A 297:138

69. Rouleau B et al (2011) Characterization at a local scale of a laser-shock peened aluminum alloy surface. Appl Surf Sci 257:7195

70. Arun Prakash N et al (2010) Microstructural evolution and mechanical properties of oil jet peened aluminium alloy, AA6063-T6. Mater Des 31:4066

71. Fribourg G et al (2011) Microstructure modifications induced by a laser surface treatment in an AA7449 aluminium alloy. Mater Sci Eng A 528:2736

72. Ren XD et al (2010) Effects of laser shock processing on 00Cr12 mechanical properties in the temperature range from 25 to 600. Appl Surf Sci 257:1712–1715

73. Ren XD et al (2013) Metallographic structure evolution of 6061-T651 aluminum alloy processed by laser shock peening: effect of tempering at the elevated temperatures. Surf Coat Technol 221:111–117

74. Prasada Rao AK et al (2004) Effect of Grain Refinement on Wear Properties of Al and Al-7Si Alloy. Wear 257:148–153

75. Juijerm P, Altenberger I (2006) Residual stress relaxation of deep-rolled Al–Mg–Si–Cu alloy during cyclic loading at elevated temperatures. Scr Mater 55:1111

76. Huang J et al (2012) Mater Eng Perform 21:915

77. Gomez-Rosas G et al (2005) High level compressive residual stresses produced in aluminum alloys by laser shock processing. Appl Surf Sci 252:883

78. Ren XD et al (2009) Influence of compressive stress on stress intensity factor of hole-edge crack by high strain rate laser shock processing. Mater Des 30:3512

79. Hu HE et al (2008) Microstructure Characterization of 7050 Aluminum Alloy During Dynamic Recrystallization and Dynamic Recovery. Mater Charact 59:1185

80. Galiyev A et al (2003) Continuous dynamic recrystallization in magnesium alloys. Mater Sci Forum 419–422:509

81. Humphreys FJ, Hatherly M (1995) Recrystallization and Related Annealing Phenomena. Elsevier Science Ltd., Oxford, p 235

82. Montross CS, Wei T, Ye L (2002) Laser shock processing and its effects on microstructure and properties of metal alloys:a review. Int J Fatigue 24:1021–1036

83. Ren XD et al (2013) Dislocation polymorphism transformation of 6061-T651 aluminum alloy processed by laser shock processing: effect of tempering at the elevated temperatures. Mater Sci Eng A 578:96–102

84. Nalla RK et al (2003) On the influence of mechanical surface treatments-deep rolling and laser shock peening-on the fatigue behavior of Ti–6Al–4V at ambient and elevated temperatures. Mater Sci Eng A 355:216–225

85. He TT et al (2011) Microstructure and hardness of laser shockedultra-fine-grained aluminum. J Mater Sci Technol 27:793–796
86. Xiong Y et al (2011) Rare Met Mater Eng 40:176–185
87. Hu YX, Grandhi RV (2012) Efficient numerical prediction of residual stress and deformation for large-scale laser shock processing using the eigenstrain methodology. Surf Coat Technol 206:3374–3385
88. Peyre P et al (1996) Laser shock processing of aluminium alloys, application to high cycle fatigue behaviour. Mater Sci Eng A 210:102–113
89. Lee WS et al (2011) Dynamic mechanical behaviour and dislocation substructure evolution of Inconel 718 over wide temperature range. Mater Sci Eng A 528:6279–6286
90. Harold L, Michael RH (2008) The effects of laser peening on high-cycle fatigue in 7085-T7651 aluminum alloy. Mater Sci Eng A 477(1–2):208–213
91. Panigrahi SK, Jayaganthan R (2011) Influence of solutes and second phase particles on work hardening behavior of Al 6063 alloy processed by cryorolling. Mater Sci Eng A 528:3147–3160
92. Buha J, Lumley RN, Crosky AG (2007) Secondary precipitation in an Al–Mg–Si–Cu alloy. Acta Mater 55:3015–3024

Chapter 4
Influence of LSP on Stress Intensity Factor of Hole-Edge Crack

Abstract This chapter focuses on the effects of the compressive residual stresses generated due to laser shock processing (LSP) on the stress intensity factor (SIF) of a through-the-thickness radial crack at the edge of the circular hole, the effects of laser shock peening on the fatigue crack initiation and propagation of 7050-T7451 aluminum alloy, and a new kind of statistical data model which described the fatigue cracking growth with limited data and the effects of the reliability and the confidence level to the fracture growth. Many materials have displayed pronounced improvements in fatigue life after LSP. It has shown that LSP treatment improves the materials mechanical properties, fatigue resistance, foreign object damage (FOD), and fatigue life.

4.1 Introduction

Laser shock processing (LSP) is a treatment for improving surface fatigue intensity of metallic materials in which residual compressive stress is mechanically produced into the surface. It is an interaction between high-energy laser and material during a very short period of time, which has been proved to be a non-conventional surface mechanical treatment used to improve resistance to the rotating bending fatigue strength of the material. LSP produces extensive plastic deformation in the metal, and fretting fatigue of high performance components in the structural and aircraft industries was improved when the peak pressure of the laser shock wave is greater than the dynamic yield strength of the material. Many materials have displayed pronounced improvements in fatigue life after LSP. It has shown that LSP treatment improves the materials mechanical property, fatigue resistance, foreign objects damage (FOD), and fatigue life.

Then, theoretical analysis work which confirmed the effect of the compressive residual stresses generated by LSP on the stress intensity factor (SIF) was put forward, showing the influence of compressive stress on the crack's SIF after LSP. However, few studies have mentioned the effects of LSP on SIF of hole-edge crack in details.

© Springer-Verlag Berlin Heidelberg 2015
X. Ren, *Laser Shocking Nano-Crystallization and High-Temperature Modification Technology*, DOI 10.1007/978-3-662-46444-1_4

The objective of this chapter was to examine the effect of LSP on SIF, analyzed SIF changing on the hole-crack subject to LSP, and influence of compressive stress. Giving a model for reliability and confidence level in fatigue statistical calculation, the effects of SIF on 7050-T7451 Al alloy by high strain rate LSP were investigated.

4.2 Stress Intensity Factor Changing on the Hole Crack Subject to Laser Shock Processing and Influence of Compressive Stress

Abstract The effect of the compressive residual stresses generated due to laser shock processing (LSP) on the stress intensity factor (SIF) of a through-the-thickness radial crack at the edge of the circular hole was investigated. The relationship between the SIF and the residual stress was determined on the basis of the weight function theory in fracture mechanics crack-propagation characteristics for such cracks subjected to the combination of the applied stress and residual stress were discussed. The influence of compressive stress by laser shock of high strain rate on SIF was deduced by means of slice synthesis model from the calculation of 3D crack tip SIF. According to the computational solution, the SIF subjecting to the modification of impact zone was smaller than that without modification when the inhibiting effect was imposed by the laser shock plastic zone on the crack development. This section aims to probe into the influence of high power, short pulse laser shock on the SIF of crack tip in metal plastic deformation zone by focusing on the inhibiting effect of residual stress produced on the crack development in the plastic zone. This shows that the compressive residual stress could lead to the decrement of the SIF. Moreover, the number of the laser shocks had an important influence on the SIF.

4.2.1 Introduction

In the past six decades, shot peening has been one of the most effective and widely used methods to introduce the compressive residual stresses into the surface of metals to improve their fatigue performance [1]. Although shot peening is relatively inexpensive and can be used on large or small areas as required, this process has its limitations. The compressive residual stresses were limited in depth, usually not exceeding 0.25 mm in soft metals such as aluminum alloys and less in harder metals. The ability of a pulsed laser beam to generate shock waves was first recognized in the early 1960s [2]. Through the subsequent efforts, LSP is emerging as an alternative technology to conventional shot peening processes for improving fatigue, corrosion, and wear resistance of materials. The effect of LSP on the material is mainly achieved through the mechanical effect produced by the shock wave, not a thermal effect from heating of the surface by the laser beam. It is well suited for precisely controlled treatment of localized fatigue critical areas, such as holes, notches, fillets, and welds [3]. The beneficial effects of LSP on surface roughness, distortion, residual stress, and hardness of materials have been concentrated due to the microstructural change to dislocation, twin crystals, grain coarsening, etc. [4–12]. The depth of the compressive stress layer near the material surface produced through LSP is usually typical in an order of magnitude greater millimeter [13, 14]. The strength of the shock wave induced by LSP has so extremely high strength that it can reach to several G Pascal or even tens of G Pascal Pa [15–17], while its impulse width is only tens of nanosecond, leading to sufficiently high strain rate (10^6–10^7 s^{-1}) near the surface of the solid materials [18]. Such high strain rate makes the solid materials exhibit some special mechanical, physical, or chemical properties [19]. The researches on the optical–acoustic and optical–mechanics coupling effects in the laser shock wave promote the formation of a cross-subject, namely photo-mechanics [20]. The residual stress has an important influence on the cracking and fatigue performance of the components. For the notched specimens, the superposition principle is often used to determine the stress intensity factor (SIF) due to the applied stress combined with the residual stress.

Phenomena such as dislocation, crystallization, and twin crystal may take place to the organization of structures of metallic and alloy materials after laser shock processing, which may probably incur significant variation to the surface roughness, deflection, residual stress, and rigidity on the material [21–24]. In particular, the compressive residual stress produced by laser shock processing on the cracking surface can reduce the tensile stress under alternating load to extend the crack initiation life [25–28]. Meanwhile, existence of compressive residual stress may cause the crack closure to incur the reduction of effective driving force for the fatigue crack growth, which is favorable for the extension of fatigue crack growth life and significant improvement of fatigue performance of the material [29, 30]. SIF serves as an important parameter in fracture mechanics, which can provide numerous well-established approaches and reliable results for 2D issues [31–33]. In recent years, the solution to 3D weight function of SIF has drawn increasing attentions. However, there are only few numerical solutions to the SIF available for finite bodies of 3D crack [34–37]. There is solution to energy difference ratio of semi-analytical and semi-engineering [38], which can be used to deduce the closed solution containing SIF of finite bodies of non-through crack and energy difference ratio, and obtain the results identical to those acquired by means of other accepted approaches. Nevertheless, it is difficult to obtain the variation of SIF along the crack as the numerical integration is required in the calculation process. Furthermore, as the solution to these SIFs requires more machine hours and manpower, it is unable to satisfy the requirements for damaged tolerance and durability design. With regard to non-through crack surface after LSP, the residual stress may witness significant variation under the instant laser shock, which may further incur the variation of crack surface SIF along the crack front. Therefore, its solution process is more complicated than the through crack issue. There are many geometrical parameters involved with the 3D issue (such as shape ratio, ratio between crack depth as well as aperture and sheet thickness). Moreover, there also exists a certain plastic zone at the crack tip when the metallic material is under laser processing, which has a decisive action on the growth, expansion, and loss of stability of the crack. Laser shock with features such as high energy, pressure, strain rate, and short pulse may produce significant residual compressive stress on the crack surface.

The aim of this section was to address the effect of the compressive residual stress due to LSP on the SIF of the crack near the hole under superimposed mechanical load and to probe theoretically into the influence of laser shock on the fatigue crack on the material surface for the purpose of establishing the expression of SIF on the 3D non-through crack surface in the process of laser impact. A weight function method was used to investigate the relationship between the SIF and the residual stress. The effective SIF was used to investigate the fracture of the specimen. The effect of the processing parameters on the crack propagation was also discussed. The 2D weight function is integrated with slice synthesis model in order to probe into the influence of compressive stress effect of plastic deformation zone on SIF of 3D non-through hole-edge crack by focusing on the inhibiting effect of residual stress produced by the laser shock on the crack development in plastic zone, and it is expected to find a solution to the growth rate of material surface crack and life prediction in case of plastic deformation of high strain rate.

4.2.2 SIF Formula

Different analytical models have been developed to predict the residual stresses induced by LSP at the material surface [39, 40]. In LSP, the shock wave propagated one-dimensionally into the elastic-perfectly plastic metal half space and the edge-effects were avoided, as shown in Fig. 4.1. If the shock-induced deformation is considered to be uniaxial and planar, the pressure pulse is considered to be uniform in space, and the materials obey Von Mises yield criterion. The plastically affected

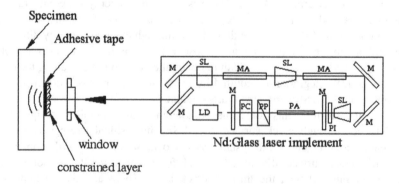

Fig. 4.1 Schematic of the LSP. Reprinted from Ref. [41], Copyright 2009, with permission from Elsevier

depth near the material surface, L_p, and the maximum surface residual stress δ_{surf} can be expressed as [40],

$$L_p = \frac{C_{el} C_{pl} \tau}{C_{el} - C_{pl}} \tag{4.1}$$

$$\delta_{surf} = -\frac{P}{2\left(1 + \frac{\lambda}{2\mu}\right)} \left[1 - \frac{4\sqrt{2}}{\pi r} \left(1 + \frac{C_{el} C_{pl} \tau}{C_{el} - C_{pl}} \right) \right] \tag{4.2}$$

where C_{el} and C_{pl} are the elastic and plastic shock wave velocities, respectively, τ is the pressure pulse duration, λ and μ are, respectively, the elastic Lame's constants of the target, γ is the radius of the impact, P is the shock wave pressure, and P is Poisson's ratio. These equations are only valid when the shock wave pressure is greater than twice the hug omit elastic limit (HEL) of the material [1].

The compressive residual stresses due to LSP lead to the variation of the SIF, K, and the surface weight function of the notched or pre-cracked specimen. Considering one diametrically radial crack emanating from the edge of a circular hole in a rectangular plate of finite width, as shown in Fig. 4.2, the SIF, K, can be calculated

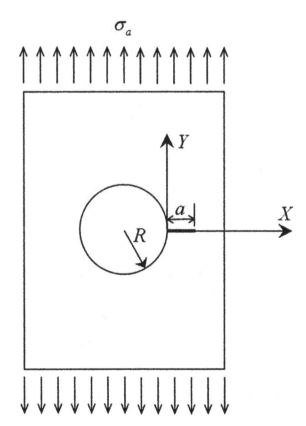

Fig. 4.2 Schematic of the radial crack emanating from the edge of a circular hole in a rectangular plate of finite width. Reprinted from Ref. [41], Copyright 2009, with permission from Elsevier

by integrating the product of the modified weight function and the nominal stress distribution [42].

$$K = \sigma\sqrt{\pi a} \int_0^a \frac{\sigma(x)}{\sigma} \frac{m(\alpha, x)}{\sqrt{\pi a}} dx \ (x = X/W, \ \alpha = a/W) \tag{4.3}$$

where $\sigma(x)$ is the stress applied on the cracking body along the crack line, σ denotes a reference stress, $m(\alpha, x)$ is the weight function, a is the crack length in the x direction, X is the coordinate along the crack direction, W is the characteristic dimension of the cracking body, and f is the geometrical factor. For the specimen with a hole, the characteristic dimension of the cracking body is assumed to be the radius of the hole [43]. For the edge crack, the weight function $m(\alpha, x)$ can be expressed as

$$m(\alpha, x) = \frac{1}{\sqrt{2\pi a}} \sum_{i=1}^{j+1} \beta_i(\alpha) \left(1 - \frac{x}{\alpha}\right)^{j=3/2} \tag{4.4}$$

where $\beta_i(\alpha) = \{\alpha F_{i=1}(\alpha) + 1/2 \times [(2i - 10F_i(\alpha) - (2i - 5)F_{i=1}(\alpha)]\}/f_r(\alpha)$ is a function expression. For a two-dimensional crack, taking $j = 2$, $i = 1, 2$, and 3, $F_i(\alpha)$ can be, respectively, expressed as [42],

$$F_1(\alpha) = 4f_r(\alpha) \tag{4.5a}$$

$$F_2(\alpha) = \left[\sqrt{2\pi}\varphi(\alpha) - E_1(\alpha)F_1(\alpha)/E_2(\alpha)\right] \tag{4.5b}$$

$$E_j(\alpha) = \sum_{m=0}^{M} \frac{2^{m+1}mS_m\alpha^m}{\prod_{k=0}^{m}(1 + 2j + 2k)}, \ j = 1, 2 \tag{4.5c}$$

where $fr(\alpha)$ is the geometrical factor and can be defined as $f_r(\alpha) = K_r(\alpha)/a\sqrt{\pi a} \ S_m$ is the polynomial modulus of the refer loading $\sigma_r(x)$, i.e., $\sigma_r(x) = \sum_{m=0}^{M} S_m x^m$, and $\varphi(a)$ can be defined as [42],

$$\phi(\alpha) = \int_0^a Sf_r^2(S)dS \tag{4.6}$$

If $f_r(\alpha)$ is defined as a polynomial formula, i.e., $f_r(\alpha) = \sum_{n=0}^{N} \lambda_n\alpha^n$, $\varphi(\alpha)$ could be rewritten as [25],

$$\phi(\alpha) = \sum_{i=0}^{N} \sum_{j=0}^{N} \frac{\lambda_i\lambda_j}{i+j+2}\alpha^{i+j} \tag{4.7}$$

Combination of the Eq. (4.4) and the Eq. (4.5a–c), the weight function [44], $m(\alpha, x)$, can be determined. Substituting the solution of $m(\alpha, x)$ into Eq. (4.3), the relationship between the SIF, K, and the randomly applied stress on the surface crack body $\delta(x)$ can be expressed as

$$K = \sigma \frac{1}{\sqrt{2\pi a}} \int_0^a \frac{\sigma(x)}{\sigma}^{j+1} \sum_{i=1} \beta_i(\alpha) \left(1 - \frac{x}{\alpha}\right)^{i-3/2} dx \qquad (4.8)$$

Prior to applied stress, the residual stress fields will be generated around the hole and along the cracking face, in such a case, the residual SIF, K_y, should be expressed as,

$$K_\gamma = \int_0^a \sigma_r(x)m(x,a)dx \qquad (4.9)$$

where $\sigma_y(x)$ is the residual stress distribution near the notch after LSP and K_r is the SIF due to the residual stress. For an edge crack, the weight functions at the bottom of the crack, A, and crack surface dot, B, can be determined by using Shen–Glinka model [34] (M_{1A}, M_{2A}, M_{3A}, M_{1B}, M_{2B}, and M_{3B} have been obtained by Ref. [45]),

$$m_A(x,a) = \frac{2}{\sqrt{2\pi(a-x)}} \left[1 + M_{1A}\left[1 - \frac{x}{a}\right]^{1/2} + M_{2A}\left[1 - \frac{x}{a}\right] + M_{3A}\left[1 - \frac{x}{a}\right]^{3/2}\right]$$

$$(4.10a)$$

$$m_B(x,a) = \frac{2}{\sqrt{\pi x}} \left[1 + M_{1B}\left[\frac{x}{a}\right]^{1/2} + M_{2B}\left[\frac{x}{a}\right] + M_{3B}\left[\frac{x}{a}\right]^{3/2}\right] \qquad (4.10b)$$

Hence, the SIFs at the bottom and surface of the crack can be, respectively, expressed as

$$K_A = \int_0^a \sigma(x)m_A(x,a)dx \qquad (4.11a)$$

$$K_B = \int_0^a \sigma(x)m_B(x,a)dx \qquad (4.11b)$$

When the specimen with a hole is subjected to the residual stress combined with the externally applied stress, the overlapping principle was used to determine the effective SIF, K_{eff}, of the crack. In such a case, the effective SIF should be composed of two parts, namely the SIF due to the applied stress, K_a, and that due to the residual stress, K_y, i.e., [17],

$$K_{eff} = K_a + K_r \qquad (4.12)$$

Table 4.1 The mechanical properties of the 2024-T62 alloy

Material	Yield strength (MPa, 25 °C)	Tensile strength (MPa, 25 °C)	Elongation percentage (%)
2024-T62	465	340	20

4.2.3 Experimental Procedures

Al specimens with a hole were manufactured by 2024-T62 alloy, which is commonly used in aero-engine fan blades, wheel hub, and airscrew. The radius of the hole R is 1.5 mm. The mechanical properties of this alloy are listed in Table 4.1.

LSP was conducted using a confined plasma configuration. A thin layer of 3-M Scotch aluminum foil adhesive tape was used as the energy absorbing layer. A thin water tamping layer was used as the plasma confinement layer. The region including the hole was shocked and nearly pure mechanical effects are induced, as schematically shown in Fig. 4.3. LSP was performed by high-power Nd:glass laser implement using different power densities and number of laser shocks to investigate the effect of the processing parameters used in LSP on the residual stress distribution. The wavelength of the laser beam is 1.054 μm; the laser pulse duration and energy are 20 ns and 35 J, respectively. The laser beam spot size was maintained at 7 mm. The residual stress near the edge of the hole was measured using the X-350A X-ray diffraction technique. The residual stress distributions along the radial direction at the surface and depth direction are determined.

4.2.4 Analysis Model

4.2.4.1 Residual Stresses

LSP in a confined geometry was distinctly different from direct ablation because the coating layer was vaporized in the case of confined geometry, whereas the workpiece itself was vaporized in the case of direct ablation. The collisions between the vaporized particles in the vaporized layer must be considered in the case of confined geometry, and the recoil and plasma pressures were of interest to calculate the pressure on the substrate surface.

The recoil pressure was determined using an approach similar to the case of direct ablation. A correction factor f_c was introduced to account for the collision of the particles in the confined geometry, and the coating layer was assumed to exist mainly in the liquid phase. The plasma pressure was calculated numerically from an energy balance for the plasma plume accounting for phase changes during the plasma formation. The radial and axial stresses could be written as [46],

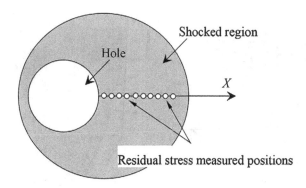

Fig. 4.3 Schematic of the region including the hole was shocked and nearly pure mechanical effects are induced. Reprinted from Ref. [41], Copyright 2009, with permission from Elsevier

$$\sigma_{rr} = \frac{v}{1 - v} \sigma_{zz} \tag{4.13}$$

$$\sigma_{zz} = \rho_1 k_B \frac{T^*(r, z = 0, t)}{M} \left((1.665 f_c)^2 \frac{T^*(r, z = 0, t)}{2\pi} + N_A \left(1 + \frac{n_e}{n} \right) T_p \right) \tag{4.14}$$

The correction factor f_c was introduced to account for the energy loss due to molecular and atomic collisions within the plasma. For the numerical evaluation, it was assumed that 15 % of the thermal energy in the plasma plume was lost due to internal collisions, i.e., $f_c = 0.85$. The numerical evaluations would be carried out for average pulse energies of 20 J, laser beam radius was 5 mm, and pulse width was 20 ns, as shown in Table 4.2 (Fig. 4.4)

The magnitude of residual stress after laser shock was positively correlated with pulse energies, with the increment of power density, the peak pressure of the shock wave increased. And the magnitude of peak pressure had a direct influence on the residual stresses, as shown in Fig. 4.5. From 2 to 2.5HEL, the residual stresses of the target gave the best result. Increasing laser power density, large range and high amplitude of stress deficiency would easily be formed on the shocked area.

Table 4.2 Laser technical parameters

Laser wave (μm)	Laser energy (J)	Laser pulse width (FWHM) (ns)	Exporting stability (%)	System ASE energy (J)	Peak value power (W)	Repetition rate (Hz)
1.054	22	20	≤±10	≤0.1	≥0.5 × 10⁹	0.5

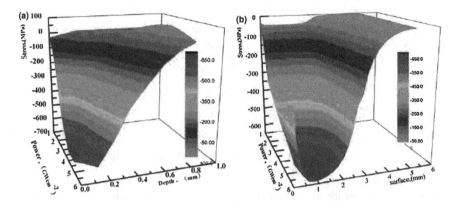

4.2.4.2 3D Stress Intensity Factor

If the rectangular sheet with hole-edge crack as indicated in Fig. 4.5 is divided with two groups of slices of uprightness angle (Fig. 4.6), the slice in parallel with the ellipse x axis should be taken as the basic slice or the transverse slice, whereas that in parallel with z axis should be taken as the leaf spring or vertical slice. The area outside of the slice zone and hole has constraint force on the cracking slices, which was called constraint area. The basic slice is in possession of the same elastic modulus E and external load as the cracking body, and the coupling action of shearing stress between two adjacent slices was simulated by the spring pressure P (x, z) imposed on the crack surface. Elastic modulus of leaf spring merely under the reverse spring force, E_S, was identical with that imposed on the basic slice. SIF

Fig. 4.5 Scheme of the rectangular sheet with hole-edge crack. Reprinted from Ref. [47], Copyright 2009, with permission from Elsevier

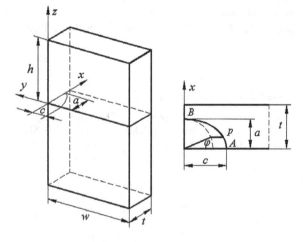

Fig. 4.6 Analysis model of slice synthesis (hole-edge surface crack). Reprinted from Ref. [47], Copyright 2009, with permission from Elsevier

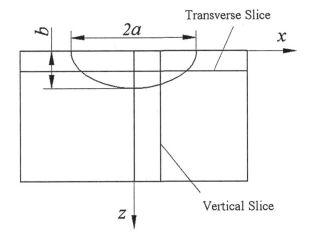

at any point of 3D crack front φ could be obtained through the composition of that of 2D cracking slices K_a and K_c, of which, the formula is as follows [48],

$$K(\varphi) = \frac{(-1)^n}{1 - \eta^2} \left\{ K_a^4(a_x) + \left[\frac{E}{E_s} K_c(c_y) \right]^4 \right\}^{1/4} \varphi \qquad (4.15)$$

where $\eta = v(\varphi < \pi/2)$. $n = 1(K_a \leq 0$ or $K_c \leq 0)$, $n = 2(K_a > 0$ and $K_c > 0)$. φ is the parameter angle of crack, definition of slice crack length a_x and c_y is as shown in Fig. 4.3, m refers to Poisson's ratio. SIF of the slices K_a and K_c could be obtained based on the 2D weight function theory

$$K_a(a_x) = \int_0^{a_x} [\sigma(x, y) - P(x, y)] m_a(a_x, y) dy \qquad (4.16a)$$

$$K_c(c_y) = \int_0^{c_y} P(x, y) m_c(c_y, x) dx \qquad (4.16b)$$

where $\sigma(x, y)$ represents the distribution of residual stress on crack surface, $P(x, y)$ refers to the spring force which could be determined based on the compatibility conditions for displacement of two groups of slices.

Slice displacement could be calculated by taking the arithmetic product of SIF and weight function as the integral [49]

$$V_a(a_x, y) = \frac{1}{E_a} \int_0^{c_y} K_a(\xi) m_a\left(\frac{\xi}{t}, \frac{y}{t}\right) d\xi \qquad (4.17a)$$

$$V_c(c_y, x) = \frac{1}{E_a} \int_0^{c_y} K_c(\xi) m_c\left(\frac{\xi}{r}, \frac{x}{r}\right) d\xi \tag{4.17b}$$

when $E_a = E$, $E_c = E_S$, and

$$\frac{E_S}{E} = \left(\frac{\varphi}{1 - v^2} - 1\right)\frac{c}{a}, \frac{a}{c} \leq 1 \tag{4.18a}$$

$$\frac{E_S}{E} = \frac{\varphi}{1 - v^2} - \frac{c}{a}, \frac{a}{c} > 1 \tag{4.18b}$$

where φ represents the second kind complete elliptic integral, which could be calculated accurately with the following formula [50],

$$\varphi = \sqrt{Q} = \left[1 + 1.464\left(\frac{a}{c}\right)^{1.65}\right]^{1/2}, \frac{a}{c} \leq 1 \tag{4.19a}$$

$$\varphi = \sqrt{Q} = \left[1 + 1.464\left(\frac{c}{a}\right)^{1.65}\right]^{1/2}, \frac{a}{c} > 1 \tag{4.19b}$$

If the displacement of two groups of slices was identical, namely $V_a = V_c$, the following equation could be obtained from formulae (4.16a, b) and (4.17a, b)

$$\int_y^{a_x} \int_0^{\xi} [\sigma(x, y) - P(x, y)] m_a\left(\frac{\xi}{t}, \frac{y}{t}\right) d\xi = \frac{E_a}{E} \int_x^{c_y} \int_0^{\xi} P(x, y) m_c\left(\frac{\xi}{r}, \frac{x}{r}\right) dx m_c\left(\frac{\xi}{r}, \frac{x}{r}\right) d\xi \tag{4.20}$$

whereas spring force could be expressed as [47]

$$P(x, y) = \sigma[\alpha_z + \alpha_2\left(\frac{x}{c}\right)^{1/3} + \alpha_3\left(\frac{y}{a}\right)^{2/3} + \alpha_4\left(\frac{x}{c}, \frac{y}{a}\right)^{1/3} + \alpha_5\left(\frac{x}{c}\right) + \alpha_6\left(\frac{y}{a}\right)$$
$$+ \alpha_7\left(\frac{x}{c}, \frac{y}{a}\right) + \alpha_8\left(\frac{x}{c}\right)^2 + \alpha_9\left(\frac{y}{a}\right)^2 + \alpha_{10}\left(\frac{x}{c}, \frac{y}{a}\right)^2 + \alpha_{11}\left(\frac{x}{c}\right)^3 + \alpha_{12}\left(\frac{y}{a}\right)^3$$
$$+ \alpha_{13}\left(\frac{x}{c}\right)^4 + \alpha_{12}\left(\frac{y}{a}\right)^4] \tag{4.21}$$

where σ is the reference stress in formula (4.20), and it is applicable to substitute formula (4.21) into Eq. (4.20) to obtain the multilinker equation as related to the undetermined coefficient α_i. We find a solution to this equation by means of regression of multilinker equation to obtain the coefficient α_i for further determination of spring force $P(x, y)$. We obtain K_a and K_c via formula (4.17a, b) before further obtaining SIF $K(\varphi)$ at each point of 3D crack front via formula (4.16a, b).

It is applicable to follow the approaches for settlement of 2D issue when calculating the SIF of 3D crack under the laser impulsive load. In other words, the distribution of residual compressive stress following the laser shock on the crack surface could be expressed in a polynomial way,

$$\sigma(x)/\sigma = \sum_{n=0}^{N} A_n(x/r)^n \tag{4.22}$$

With regard to stress distribution in the form of exponential function $(x/r)^n$, the corresponding geometric factor could be marked as

$$f_{dpn} = K_{dpn}/\left(\sigma\sqrt{\pi a/Q}\right) \tag{4.23}$$

f_{dpn} could be listed into a form to facilitate the engineering application. It is applicable to easily calculate the SIF under the polynomial distributing load,

$$K = f\sigma\sqrt{\pi a/Q} \tag{4.24}$$

$$f = \sum A_n f_{dpn} \tag{4.25}$$

For instance, a specific residual stress field adjacent to the hole wall area could be expressed as [48]

$$\sigma(x)/\sigma = \sum_{n=0}^{3} A_n(x/r)^n, \ x/r \le 1 \tag{4.26}$$

In formula (4.14), polynomial coefficient could be expressed as $A_0 = -1.5$, $A_1 = 3.32$, $A_2 = -2.16$, and $A_3 = 0.63$. It was only necessary to substitute these coefficients into formula (4.12) for summation by means of arithmetic product with the corresponding f_{dpn} to rapidly obtain the SIF of hole-edge dual surface crack under the action of residual stress field of laser impact.

Viewing formula (4.14), the dimensionless SIF value (marked as $-F_{dc}$) corresponding to various a/t in case of $r/t = 1$, $a/c = 1$ could be deduced. This approach for disposal is also applicable to the calculation of SIF of other types of non-through cracks under complicated load actions.

4.2.4.3 Correction of 3D Intensity Factor in Shocking Zone

Some scholars preferred to incorporate the influence of plastic zone on crack properties into the calculation of SIF based on Irwin's theory on correction of plastic zone [51]. This approach was known as equivalent crack method, which would incur the amplification of SIF in most cases. However, there always existed a plastic zone (namely, the yield zone) on the crack tip of most of substantial metallic materials, which were called small-scale yielding when the size of plastic zone was not so big. At this point, most area on the crack tip was known as the elastic range. Residual stress was introduced on the crack development in the plastic zone. It was also expected to provide a formula for the calculation of SIF and computation approaches in consideration of the correction of effect of plastic zone in inhibiting the crack growth (Fig. 4.7).

As indicated in Fig. 4.8, $2a$ refers to the crack length, whereas b represents the depth of surface central crack. The size of plastic stress deformation zone on the crack tip is ρ_A and ρ_B. According to fracture mechanics, the following formula could be obtained in case of small-scale yield [35],

$$\rho_A = \frac{\pi}{8}\left[\frac{K_A}{\alpha_A \sigma_{YS}}\right]^2 \qquad (4.27a)$$

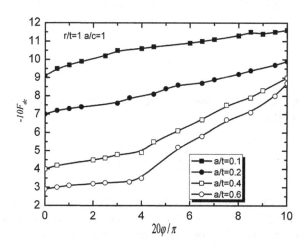

Fig. 4.7 Dimensionless crack length verses effective stress intensity factor of the hole symmetry surface crack. Reprinted from Ref. [47], Copyright 2009, with permission from Elsevier

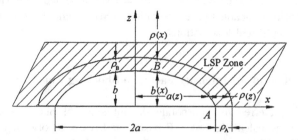

Fig. 4.8 Computational model of the surface center crack on laser shock zone. Reprinted from Ref. [47], Copyright 2009, with permission from Elsevier

$$\rho_B = \frac{\pi}{8}\left[\frac{K_B}{\alpha_B \sigma_{YS}}\right]^2 \tag{4.27b}$$

Displacement at any point (x, z) on the crack surface under external load and load imposed by laser incurring residual stress could be expressed, respectively, as follows [51],

$$\omega_a(x, z) = \frac{1}{E} \int_x^{a(z)+\rho(z)} \int_{-\xi}^{\xi} \{\sigma(u, z) - p(u, z)\}g_a(u, \xi)du\,g_a(x, \xi)d\xi$$

$$- \frac{1}{E} \int_x^{a(z)+\rho(z)} \int_{u(z)}^{\xi} \alpha_a \sigma_{YS} g_a(x, \xi)d\xi du \tag{4.28}$$

$$\omega_a(x, z) = \frac{1}{E_s} \int_z^{b(x)+\rho(x)} \int_{-\xi}^{\xi} p(x, v)g_b(v, \xi)dv\,g_b(z, \xi)d\xi$$

$$- \frac{1}{E} \int_z^{b(x)+\rho(x)} \int_{b(x)}^{\xi} \alpha_b \sigma_{0S} g_b(v, \xi)g_b(z, \xi)d\xi dv \tag{4.29}$$

In formula (4.28), σ_{0S} refers to the equivalent yield limit of leaf spring. It is applicable to calculate the 3D ellipse crack and define r_{0S} as the leaf spring by means of slice synthesis model. The computational formula is stated as [50],

$$\frac{\sigma_{0s}}{\sigma_{Ys}} = \frac{E_s}{E} = \left[\frac{\varphi}{1 - v^2} - 1\right]\frac{b}{a}, \frac{b}{a} \leq 1 \tag{4.30a}$$

$$\frac{\sigma_{0s}}{\sigma_{Ys}} = \frac{E_s}{E} = \frac{\varphi}{1 - v^2} - \frac{b}{a}, \frac{b}{a} > 1 \tag{4.30b}$$

φ is the complete elliptic integral of the second kind. α_a and α_b represent the constant corresponding to the basic slice and leaf spring on any location, respectively, of which, the value varied in line with the changing positions. In terms of treatment to linear variation, it is expressed as [51],

$$\alpha_a = \frac{\alpha_B - \alpha_A}{\pi/2}\arccos\left[\frac{a(z)}{a}\right] + \alpha_A \tag{4.31a}$$

$$\alpha_a = \frac{\alpha_B - \alpha_A}{\pi/2}\arcsin\left[\frac{a(z)}{a}\right] + \alpha_A \tag{4.31b}$$

According to the laser deformation zone as indicated in Fig. 4.8, deformation zone induced by LSP is shown as

$$\rho(z) = (a + \rho_A)\sqrt{1 - \frac{z^2}{(b + \rho_B)^2}} - a\sqrt{1 - \frac{x^2}{b^2}} \tag{4.32a}$$

$$\rho(x) = (b + \rho_B)\sqrt{1 - \frac{x^2}{(a + \rho_A)^2}} - b\sqrt{1 - \frac{x^2}{a^2}} \tag{4.32b}$$

Supposing the spring force $p(x, z)$ is related to the power polynomial of x and z [35],

$$p(x, z) = \sum_{i=0}^{2} \sum_{j=0}^{2} a_{ij} x^i z^j \tag{4.33}$$

The following formula could be obtained as per compatibility conditions for displacement,

$$Y(x, z) = \sum_{i=0}^{2} \sum_{j=0}^{2} \alpha_{ij} X(x, z)_{ij} \tag{4.34}$$

In formula (4.34), where

$$Y(x, z) = \int_{x}^{a(z)+\rho(z)} \int_{-\xi}^{\xi} \sigma g_a(u, \xi) du g_a(x, \xi) d\xi - \int_{x}^{a(z)+\rho(z)} \int_{a(z)}^{\xi} \alpha_a \sigma_{YS} g_a(u, \xi) g_a(x, \xi) d\xi du$$

$$+ \frac{E}{E_s} \int_{z}^{b(x)+\rho(x)} \int_{b(x)}^{\xi} \alpha_b \sigma_{0S} g_b(z, \xi) d\xi dv \tag{4.35a}$$

$$X(x, z)_{ij} = \int_{x}^{b(x)+\rho(x)} \int_{-\xi}^{\xi} u^i z^j g_a(u, \xi) du g_a(x, \xi) d\xi$$

$$+ \frac{E}{E_s} \int_{z}^{b(x)+\rho(x)} \int_{0}^{\xi} x^i v^j g_b(v, \xi) dv g_b(z, \xi) d\xi \tag{4.35b}$$

Thus, the continuous expression formula for displacement could be established. It is also applicable to make use of linear regression to obtain the coefficient α_{ij} as well as $p(x, z)$. In consideration of points A and B in the deformation zone, SIF could be expressed as follows,

$$K_{IA} = \int\limits_{-(a+\rho_A)}^{a+\rho_A} \{\sigma - p(x,0)\}g(x,a+\rho_A)_{c,c}\,\mathrm{d}x - \int\limits_{0}^{a+\rho_A} \alpha_A \sigma_{YS}g(\xi,a+\rho_A)_{c,c}\,\mathrm{d}\xi$$

(4.36a)

$$K_{IB} = \frac{E}{E_s}\int\limits_{0}^{b+\rho_B} p(0,z)g(z,b+\rho_B)_{c,c}\,\mathrm{d}z - \frac{E}{E_s}\int\limits_{0}^{b+\rho_B} \alpha_B \sigma_{0s}g(\xi,b+\rho_B)_{c,c}\,\mathrm{d}\xi \quad (4.36b)$$

It is applicable to obtain the solution to SIF in consideration of the correction to the plastic zone (hereafter referred to as corrected solution) by taking the surface central crack indicated in Fig. 4.5. When the width w, thickness t, Poisson's ratio as well as crack depth b were determined as 130, 20, 0.3, and 3.15 mm, respectively, values of a/b would be 4.999, 2.800, 1.667, 1.250, 0.9621, and $\alpha_A = \alpha_B = 1$, respectively. Meanwhile, the dimensionless SIF (hereafter referred to as initial solution) on the crack tip should be indicated in the dimensionless form on condition that ideal linearity was calculated. Curve collation map for initial and corrected solutions is shown in Fig. 4.9. From the correlation curve for the two solutions, it could be seen that SIF of corrected solution was smaller than that of the initial solution. In consideration of large dimension of plastic zone and significant influence of plastic zone, it is applicable to consider the correction of plastic zone. In other words, it was necessary to consider the correction of plastic zone when calculating the SIF of 3D crack on the thinner structure in an inclination to plane stress state.

Fig. 4.9 Curve contrast chart of the initial solution and modified solution. Reprinted from Ref. [47], Copyright 2009, with permission from Elsevier

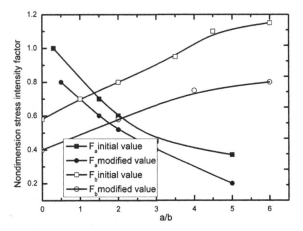

4.2.5 Results and Discussion

The measured residual stresses along the depth direction and at the surface as functions of the power density are shown in Fig. 4.10a, b, respectively. It can be seen that the power density, which is controlled by the shock wave pressure, has a significant effect on the magnitude of the residual stress due to LSP. When the power density is lower than 2 GW/cm², it only has an influence on the residual stress at the surface. At the subsurface, the residual stress is not sensitive to the power density, as shown in Fig. 4.10a. However, from Fig. 4.10a, two important insights can be found. First, the compressive residual stress at the surface increased with the power density increases. The result agrees well with the analytical result obtained from Eq. (4.2). Second, with increasing the depth, the residual stress changes from compressive to tensile to satisfy the balance of the forces due to the residual stress. Moreover, with increasing of the power density, the distance between the locations at which the compressive stress changes to tensile stress and the surface increases.

Fig. 4.10 The measured residual stresses along the depth direction and at the surfaces functions of the power density: **a** surface direction and **b** depth direction. Reprinted from Ref. [41], Copyright 2009, with permission from Elsevier

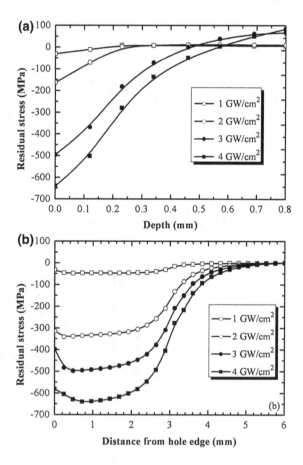

Hence, it also can be concluded that the thickness of the plastic layer increases with increasing the power density. From Fig. 4.10b, it can be seen that with increasing the distance from edge of the hole, the compressive residual stress at the surface increases and reaches to a local maximum and then decreases. When the distance from the edge of the hole is very large, the residual stress is almost to be zero.

The normalized effective SIF at the crack surface near the edge of the hole as functions of the normalized crack length with respect to two different values of the applied stress is shown in Fig. 4.11, where σ_a is the applied stress, σ_y is the yield strength of the 2024-T62 alloy, and the effective $\sqrt{R}$ SIF is expressed as $K_n = \sqrt{3}\,K_{\mathrm{eff}}/(\sigma_a + \sigma_y)$ [35]. The effective SIF is determined through Eq. (4.9–4.12) and the measured residual stresses at the power density of 3 GW/cm^2 in Fig. 4.5 is used to determine the SIF due to residual stress field. It can be seen that at a given crack length, the normalized effective SIF, K_n, decreases with increasing the applied stress, σ_a. When the applied stress is $0.3\sigma_y$, the normalized SIF is zero as the normalized crack length a $\alpha \le 2$. In such a case, the crack will be not propagated. Hence, it can be concluded that the compressive residual stress near the hole has an important influence on the radial crack propagation when the applied stress is relatively low ($\sigma_a = 0.3\sigma_y$; σ_y is the external force). However, when the applied stress is high, this influence is not obvious. All those declare that surface residual stress has a remarkable influence on the material anti-fatigue strength. Surface residual stress acts as an average residual stress. And residual compressive stress acts as a minimum average residual stress, which would enhance anti-fatigue strength; while residual tensile stress acts as a positive average residual stress, which would decrease anti-fatigue strength.

The normalized effective SIF as functions of the normalized crack length with respect to different number of laser shocks in LSP is shown in Fig. 4.12, where the applied stress is kept to be 0.15 σ_y. It can be seen that by comparing with the untreated specimen, the SIFs, K_n, of the specimens treated by LSP are relatively

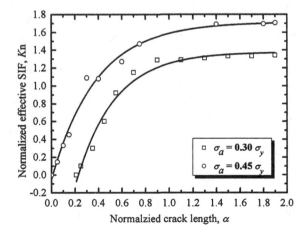

Fig. 4.11 The normalized effective SIF with respect to two different values of the applied stress. Reprinted from Ref. [41], Copyright 2009, with permission from Elsevier

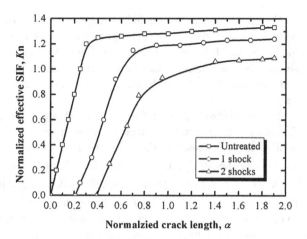

Fig. 4.12 The normalized effective SIF with respect to different number of laser shocks in LSP. Reprinted from Ref. [41], Copyright 2009, with permission from Elsevier

low, indicating that the LSP has the beneficial effect on the decrement of the growth rate of the fatigue cracks. Moreover, when the crack length, α, is given, the SIF decreases with increasing the number of the laser shocks. For the specimen treated by two laser shocks, the effective SIF is zero when $\alpha \le 0.39$, whereas for the specimen treated by one laser shock, $K_n = 0$ where $\alpha \le 0.21$. This result indicates that the growth rate of the crack decreases with increasing the number of the laser shocks used in LSP.

4.2.6 Conclusions

The relationship between the SIF and the residual stress was determined on the basis of the weight function theory in fracture mechanics. As to crack problem of the residual stress field, the effect of residual stress on SIF could be attained by weight function, and the fringe crack SIF could be derived by the outside load with the weight function after LSP. The normalized effective SIF as functions of the normalized crack length with respect to different number of laser shocks indicates that the growth rate of the crack decreases with increasing the number of the laser shocks. And the compressive residual stress near the hole has an important influence on the radial crack propagation when the applied stress is relatively low.

This section integrates the 2D weight function with slice synthesis model to determine the interaction force between slices under the compatibility conditions for displacement when calculating the SIFs of two groups of slices by means of 2D weight function, and finally obtaining the SIF value at each point of non-through crack front based on the interrelation between SIF of 3D crack and that of 2D slices.

The linear elastic fracture mechanics takes the theory of linear elasticity as the basis, which is only applicable to the pure linear elastic crack body. Elastic mechanic analysis will continue to be applicable as it is necessary to correct the influence of the plastic zone by laser shock processing. The influence of correction to the plastic zone on the 3D crack tip on the crack growth life after the inhibiting effect of residual compressive stress was produced by the laser impact based on the calculation of SIF of 3D crack tip by means of slice synthesis model. According to the computational results, SIF subjecting to correction of plastic zone was lower than that without correction of plastic zone by 11–35 % in most LSP cases.

Hence, another further work should be carried out to perform the residual stress measurements for the aluminum alloys after laser shock processing. The comparison between the analytical result obtained from the present model and the experimental result should also be made, and a more accurate stress and SIF expression should also be made.

4.3   Investigation of Stress Intensity Factor on 7050-T7451 Aluminum Alloy by High Strain Rate Laser Shock Processing

Abstract The aim of this unites was to identify the effect of laser shock peening on the fatigue crack initiation and propagation of 7050-T7451 aluminum alloy. The laser shocked specimen in which residual compressive stress is mechanically produced into the surface showed a very high dislocation density within the grains. This was evident throughout the LSP region. The spacing among the fatigue striations in the LSP region was narrow, which indicated that LSP had an obvious inhibitory action to fatigue crack initiation and growth. In contrast, the region without LSP exhibited an extremely low dislocation density. And LSP improved 7050-T7451 alloy specimens' fatigue intensity.

4.3.1   Introduction

Laser shock processing (LSP) is a competitive technology as a method of imparting compressive residual stresses to improve fatigue and corrosion properties, the principle of LSP is to use a high intensity laser and suitable overlays to generate high pressure shock waves on the surface of the workpiece. During LSP, the action on the surface and subsurface of the materials is achieved through the mechanical effect arisen from the laser shock wave.

LSP has been shown to be effective in improving the fatigue properties of 6061-T6 aluminum alloy [52], 00Cr12 alloy [53], and 2024-T3 Al alloy [54]. Micro-structural changes induced by LSP were related to the laser processing parameters [55] and the heat treatment condition of alloys [56]. The micro-structural changes lead to the improvement of the mechanical properties, such as fretting fatigue resistance [57], tensile strength [58], wear [59], and fatigue strength [60]. Most existing researches have focused on the mechanical properties improvement of aluminum alloy after LSP. Masaki et al. [61] revealed that the surface roughness, which was related to the processing parameters during LSP, had an important influence on the fatigue performance of aluminum alloy. Clauer et al. [62] investigated the effect of LSP parameters on the fatigue crack growth rate of different aluminum alloys. It was generally believed that the fatigue life improvement by laser processing could be separated into a relatively large increase in the early crack growth stage and a small increase in the later propagation stage. The influence of compressive stress effect of plastic deformation zone on stress intensity factor of the hole edge was investigated [63], and the next results show the influence of compressive stress on the 3D non-through hole-edge crack's SIF after LSP [64].

However, few studies have mentioned the effects of LSP on the hole fatigue behavior and crack propagation [65]. Residual stress causes plastic strain, leading to crack growth rate, and changes in the way extension. The objective of this work was to examine the effect of LSP on the fatigue behavior of 7050 specimens hole surface. The effects of LSP on fatigue crack behavior and residual stresses of the hole were investigated.

4.3.2 Experiments and Method

7050-T7451 aluminum alloy was chosen for this study. The compositions of the alloy are shown in Table 4.3, and the dimensions and schematic of standard stretches fatigue specimen are shown as Fig. 4.13.

According to Aerospace Material Specification (AMS) 2546 "Laser peening standard [67], LSP was performed by high power Nd:Glass laser implement." The shock waves were induced by a Q-switched repetition-rate laser with a laser beam wavelength of 1.054 mm, pulse duration and power density were 20 ns and 2.91 GW/cm^2, respectively, and laser beam spot size was maintained at 3 mm. LSP experiment was conducted with a confined plasma configuration. A thin aluminum foil adhesive tape was used as the energy absorbing layer, and a water tamping layer with a thickness of about 2 mm was used as the plasma confinement layer. The processing parameters used in LSP were shown in Table 4.4 in detail. When the power density of the laser pulse is sufficiently high, the shock waves will be induced.

Table 4.3 Chemical compositions of 7050-T7451 alloy

Composition percent %					
Trademark	Si	Fe	Cu	Mn	Cr	Zr
7050	≤0.12	≤0.15	2.0–2.6	≤0.10	1.9–2.6	≤0.9–2
	Zn	Ti	Al (single)		Al (summation)	
	5.7–6.7	≤0.06	≤0.05		≤0.15	

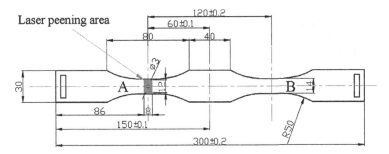

Fig. 4.13 The morphology and the size of the 7050-T7451 crack specimen. Reprinted from Ref. [66], Copyright 2013, with permission from Elsevier

Table 4.4 The processing parameters used in LSP

Laser power density (GW/cm^2)	Laser spot diameter (mm)	Pulse duration (ns)	Laser wavelength (nm)	Export stability (%)
2.91	3	20	1054	<5

The shock wave propagates into the worked target; the compressive residual stress will be generated near the target surface due to the plastic deformation.

All residual stress measurements were performed by a standard X-ray diffraction technique according to the $\sin^2\psi$-method, using $Cr–K$ radiation for stress determination. The measurements were collected at different locations across the LSP region. Depth profiles were obtained by successive electrochemical removal of the material. Sample cycle loading to investigate residual stress relaxation was performed in a MTS880 closed-loop universal testing machine. The load was cycled between 0 and 350 MPa upper limit (equivalent to 85 % of yield strength) at a stress ratio of $R = 0.1$, starting from low to high. Fraction-graphic observations were performed using standard scanning electron microscopy (SEM). In particular, the spacing of the fatigue striations on the fracture surfaces was measured on surfaces perpendicular to the incoming electron beam. All of these measurements were intended to give an indication of the local fatigue behavior after LSP.

4.3.3 Results and Discussions

4.3.3.1 SIF Effect on Fatigue Crack

The residual stress was measured for circumferential longitudinal direction and transverse direction from LSP-worked hole. There exists a large compressive residual stresses in the surface layer. The comparison of measurement results of circumferential residual stresses is presented in Fig. 4.14 and radial residual stresses in Fig. 4.15, respectively.

Fig. 4.14 Experiment results of circumferential residual stress of the 7050-T7451 crack specimen. Reprinted from Ref. [66], Copyright 2013, with permission from Elsevier

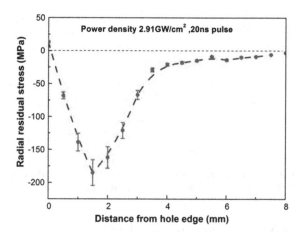

Fig. 4.15 Radial residual stress field near the expanded hole of the 7050-T7451 crack specimen. Reprinted from Ref. [66], Copyright 2013, with permission from Elsevier

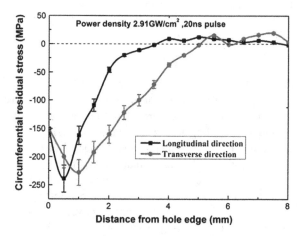

The maximum compressive residual stresses were about −250 MPa for longitudinal direction and about −220 MPa for transverse direction. The results disagreed for distances at the vicinity of the hole where it is not possible to resolve the steep stress gradients by X-ray diffraction. It is observed that the residual tangential stress varies significantly through the thickness, particularly in the region near the hole edge. High compressive stresses near the LSP-worked hole edge play an important role in arresting fatigue crack propagation. According to cold expansion techniques, the hole was reamed to final dimension after cold expansion. This process can induce the release of residual stress and redistribution of residual stress around the LSP-worked hole [68].

Residual stresses changed around the LSP-worked hole, and surface stress intensity factor K also changed, which affected the function of crack length with a given applied load. SIF could be used to quantify the magnitude of the stress singularity. The SIF may then be used to determine the likelihood of fast fracture or to infer the likely crack growth rates under fatigue-type loading. A fracture mechanics, which derivate threshold stress intensity range of the material fatigue crack expand ΔK_{th} and crack growth velocity d_a/d_n, was used to design materials' fatigue life. Fatigue life of the metallic structures was decided by ΔK and stress ratio R. The actual effect of LSP on fatigue behavior is more complex than be explained by residual stresses. It is important to point out that closure contribution at a certain value K and residual stresses producing value K at the LSP-worked hole crack tip are different and independent concepts. Residual stresses, which produce a residual SIF, K_{rs}, at the hole crack tip, could be shown as

$$K_{eff} = K_w + K_{rs} \tag{4.37}$$

where K_w is the effective values caused by the outside load; K_{rs} is caused by the residual stress.

The effective values of K based on maximum and minimum cyclic loads are given by K_{max} and K_{min}, respectively. Then

$$K_{eff\,max} = K_{W\,max} + K_{rs}; \quad K_{eff\,min} = K_{W\,min} + K_{rs} \tag{4.38}$$

The effective values of the surface SIF K_{eff} and the effective values of crack growth velocity R_{eff} could be deduced as

$$\Delta K_{eff} = K_{eff\,max} - K_{eff\,min}$$
$$R_{eff} = \frac{K_{eff\,min}}{K_{eff\,max}} \tag{4.39}$$

Residual stresses are compressive stresses at the LSP-worked hole crack tip, and K_{rs} is mathematically negative, this means ΔK_{eff} could be divided into two kinds.

1. Residual stresses effect is less than the effect of the outside stress, and the minimum effective SIF could be defined as $K_{eff} = (K_{min} - K_{rs}) > 0$, then

$$\Delta K_{eff} = (K_{max} - K_{rs}) - (K_{min} - K_{rs}) = K_{max} - K_{min} = \Delta K \tag{4.40}$$

In this case, R_{eff} decreases as K_{rs} is a negative number, namely

$$R_{eff} = \frac{K_{min} - K_{rs}}{K_{max} - K_{rs}} < \frac{K_{min}}{K_{max}} = R \tag{4.41}$$

This means that the minimum effective SIF of the crack tip unchange, while the crack growth velocity decreases as the stress ratio R decreases at the LSP-worked hole.

2. Residual stresses effect is higher than the outside stress, the minimum effective SIF could be defined as $K_{eff} = (K_{min} - K_{rs}) < 0$, and compressive residual stresses would not cause crack expanding, $K_{eff} = (K_{min} - K_{rs}) = 0$, then

$$\Delta K_{\text{eff}} = (K_{\max} - K_{\text{rs}}) - (K_{\min} - K_{\text{rs}}) = K_{\max} - K_{\text{rs}} - 0 \le \Delta K$$

$$R_{\text{eff}} = \frac{K_{\min} - K_{\text{rs}}}{K_{\max} - K_{\text{rs}}} = 0 \tag{4.42}$$

In this case, the minimum effective SIF of the LSP-worked hole crack tip decreases, while the crack growth velocity would decrease as the ΔK_{eff} decreases at the LSP-worked hole. So, it is possible $K_{\text{eff}} < 0$, and the results demonstrate that the crack-propagation life would be longer because of deep surface residual stresses. An obvious characteristic of the SIFs is that there is a minimum value for a certain length of crack. This is due to the maximum compressive tangential residual stress occurring away from the hole edge.

4.3.3.2 Residual Stress on the Fatigue Crack

Figure 4.16 shows crack-propagation behavior of the LSP-worked hole at different stress amplitudes. For non-worked specimens, cracks grow faster than LSP-worked hole specimens. The compressive residual stresses at the crack tip state that compressive residual stresses are developed at the crack tip due to the large plastic zone after the overload. Upon removing the overload, compressive stresses are developed as the elastic body surrounding the crack tip trying to squeeze the overload plastic zone back to its original size.

In the LSP-worked hole specimens, a large deformed plastic zone occurs at the crack tip by means of compressive residual stresses. In addition, the beneficial effect of the LSP is clearly observed. Surface residual stress makes the fatigue limit stress increase greatly of the LSP-worked hole, which increases the fatigue crack expanding threshold, and the LSP-worked region comes to crack hard, on the otherwise the crack grows into micro-multiplier crack, which made the crack expanding path more twist.

Fig. 4.16 Crack-propagation behavior for hole specimens at different stress amplitudes. Reprinted from Ref. [66], Copyright 2013, with permission from Elsevier

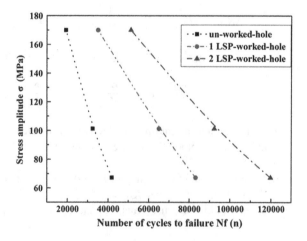

The fracture surfaces were analyzed in order to determine the crack front and establish an eventual influence of mandrel entry direction on fracture. The crack length as a function of the number of cycles was derived by processing the digital images obtained through fatigue tests. Figure 4.16 presents an example of the SEM of fatigue striations for the same crack length of $a = 0.038$ mm. Whereas the presence of the mechanical treated surface is generally considered to enhance S/N fatigue life-times primarily by inhibiting the initiation of cracks, SEM measurements of the spacing of the fatigue striations on the fracture surfaces also show a beneficial effect in slowing down the initial crack-propagation rates. Factor-graphic data acquisition was performed by five measurements in each SEM screen (as shown in Fig. 4.17a, b).

Fatigue striation patterns were clearly observed on the fracture surface of both the LSP-worked and unworked specimen. For each crack length, the mean value of measured striation spacing was calculated. In general, the striations spacing may have different values in different locations in a given screen, and striations are not completely regular.

Fig. 4.17 Fatigue striation spacing: **a** none-worked hole, crack length a = 0.038 mm; **b** LSP-worked, crack length a = 0.038 mm. Reprinted from Ref. [66], Copyright 2013, with permission from Elsevier

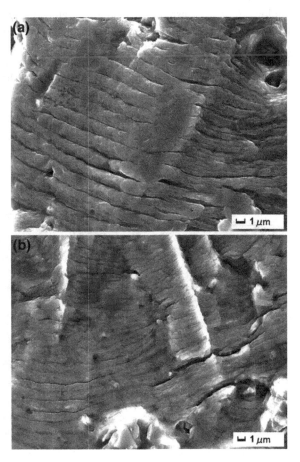

However, the spacing of the fatigue striations in the LSP-worked specimen is narrower than those observed on the unworked specimen. Figure 4.17b shows an improvement of stable crack propagation for all stress levels before fracture. The variation of the slope of the crack length curve is correlated with the values of the SIFs present at the crack tip. In fact, the compressive residual stress diminishes the effective SIFs, i.e., reduces the rate of crack growth. When the crack propagates within the tensile residual stress field, the residual SIF cumulates with the applied SIF, considerably enhancing the slope of the crack length curve. This is indicative of a slower fatigue crack growth rate. The fatigue strength of the prefabricated eyelet samples increases straightly after LSP, respectively. Additional simulations could be performed to estimate the LCF life as well [69], but this is beyond the scope and purpose of this section.

4.3.3.3 Fracture Morphology

Figure 4.18a shows the diagrams of characteristic distribution of fracture and morphology of the expansion source area separately. Crack source region locates in the hole surface away from the laser worked zone, and steady-state expansion area is the mainly mechanism of striation cycle where there are fatigue striations (FS) in every fracture surface; Some cleavage fracture morphologies appear, and a large number of secondary cracks are also generated. Figure 4.18b shows the morphology of the

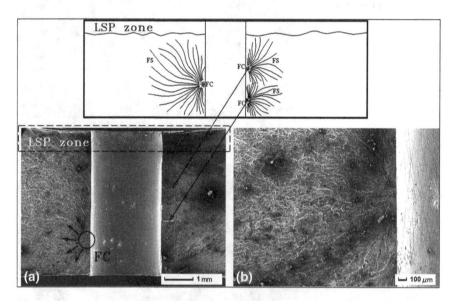

Fig. 4.18 Character distribution diagram of fracture of the LSP-worked specimens. **a** Fatigue crack from the monocarps of the specimen; **b** cleavage crystal plane and the minor striation of fatigue crack source region. Reprinted from Ref. [66], Copyright 2013, with permission from Elsevier

cleavage plane in the fatigue source region, with the fatigue cycle the stress concentration is formed in the sharp of the root of eyelet gap, and gradually cracks are formed and expand along a certain direction angle in the load direction. Fatigue crack originates in the local micro-region with strain concentration. Although the main form of plastic strain is slip, the distribution of slip is different from each other when plastic strain occurs, and the slip of cyclic plastic strain is limited to be in certain crystal grain. This kind of slip is generated in the sample surface and gradually expands into the interior. With the increase of the number of circulations, the density of slip bands increases. When a large number of slip bands are formed, the cracks are generated along the direction of the slip bands.

The steady-state expansion area is also mainly a mechanism of the belt expansion where there are some cleavage steps, and there are some small planes in the direction of vertical loading in the steps. A large number of secondary cracks are generated of the LSP-worked hole, as shown in Fig. 4.19. Cleavage planes and cleavage steps appear rugged at each lamella, and many secondary cracks appear among the lamellas and then expand to the internal material. The emergence of these secondary cracks will cost most of energy, which will also slow down the expansion of cracks [70]. Figure 4.20 shows the fracture surface morphology of the rapid expansion area, where most of cleavage planes and cleavage steps appear. There are a large number of cleavage steps and fatigue striations in the LSP-worked zone; it shows some fracture features whose stress condition is very complex (Fig. 4.20), and there are a large number of shallow dimples with instantaneous fault near the LSP-worked surface. The final fracture rate of the samples will be slowed because of the compressive residual stress, and this is presented by the material crystal characteristic, which is reflected by the dimples in Fig. 4.20. Many parallel stripe morphologies are observed in the fracture surface, and a large number of secondary cracks which are perpendicular to the main fracture surface appear at the inter-stripes, and this phenomenon is similar to Fig. 4.19.

Fig. 4.19 Typical fatigue striation in the crack expansion area of the LSP-worked specimens. Reprinted from Ref. [66], Copyright 2013, with permission from Elsevier

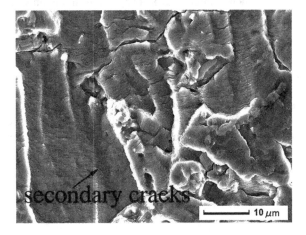

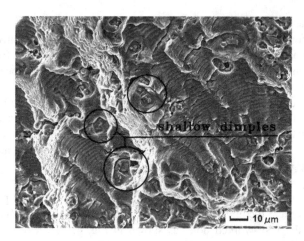

Fig. 4.20 It shows that elastic strain profiles have similar features as mean stress profiles. The peaks appearing in the plastic strain profile suggest the existence of high dislocation density in this region relative to the other places. Reprinted from Ref. [66], Copyright 2013, with permission from Elsevier

4.3.4 Conclusions

This section analyzes the influence of compressive stress on SIF of the hole edge. The effect of LSP on the hole surface fatigue behavior of 7050 specimens is also revealed.

1. Residual stresses changed in the LSP-worked zone, and surface SIF also changed. The maximum compressive residual stresses were about −250 MPa for longitudinal direction and about −220 MPa for transverse direction.
2. 7050-T7451 alloy hole crack initiation and growth are influenced by LSP. The specimens' fatigue cracks originate from the unworked area surface, the spacing among the fatigue striations of the LSP-worked hole is narrow, and the number of fatigue striations is large. As a result, the effective SIF that controls the fatigue crack growth in the LSP-worked specimen is lower than that of the unworked case.

The comparison of measurement results shows that the distance between fatigue striations is narrow and the number of striations is very large in the LSP-worked zone, which indicates that the crack expansion distance is small, and it means that LSP has an inhibition effect on the material fatigue crack initiation and expansion.

Hence, another further work should be carried out to reveal the SIF on fatigue behavior and crack propagation of the hole.

4.4 A Model for Reliability and Confidence Level in Fatigue Statistical Calculation

Abstract A new kind of statistical data model which described the fatigue cracking growth with limited data was proposed in this chapter, and the effects of the reliability and the confidence level to the fracture growth were considered. The one-sided allowance factor statistical analysis method was used to provide the prediction of the fatigue life with the confidence level and the reliability, and the effect factors were revised, which were closer to the lower limit of the matrix hundred rank values. It was found that this method gave much more accurate fatigue life prediction by analyzing the statistical data of the 7050 aluminum alloy before and after LSP and the new one-sided allowance coefficient saves more test samples in the same situation of precision. The revision coefficient would also save the experimental work load in the experiment.

4.4.1 Introduction

Stress-based fatigue analysis has been favored in many engineering practices recently, and several valuable efforts have been made. The conventional fatigue reliability method is based on the survival probability (reliability) curve [71], namely the P–S–N curve, which is reliable only when the quantity of experimental data is big and uncertainty factors that affect the fatigue performance are exposed. The conventional method mentioned the concept of confidence level [71], but it has not yet developed the random reliability S–N curve category. Whilst tests on simple components could provide some confidence, increasing understanding and prediction accuracy, while it is necessary to test full size components in their in-service environment with a representative load regime [72–74]. Recently, one-dimensional loading was carried out to take into account the probabilistic effects on the fatigue behavior by giving a probabilistic stress-number of cycle's curves [75, 76]. The weakest-link theory is the first approach treating the statistical effect in fatigue. An approach uses the Weibull model [77]. This model has been applied to bearing steel by Bomas et al. [78] and a cast iron to explain the statistical distribution on fatigue strength by Hild and Roux [79] and Chantier et al. [80] to explain the statistical distribution on fatigue strength. More recently, a combination of the concept of the weakest-link and a critical plane damage model based on a micro-plasticity analysis to describe the distributions of the fatigue limit and the fatigue life under different loadings was analyzed by Morel [81]. Hence, an alternative approach is commonly adopted in order to evaluate the reliability in engineering design, experimental, or numerical simulation, and load histories are counted, as they are in deterministic-approaches [82–85]. Then, the obtained counted cycles are statistically analyzed by means of specific procedures.

The coefficient of variation is caused to rise in a large scale, thus the request of the number of the smallest test sample is difficult to be satisfied in the short-life area frequently, the insufficient data and the uncertainty factors that result to the measure results tend to lead a possible danger. So, the concept of the one-sided allowance coefficient is needed to give lower confidence of matrix hundred rank values with the confidence level. Therefore, seeking for the fatigue reliability analysis method which considers situation of limited data with the important theory significance and the project application should be worthy to be explored. The object of this chapter is to develop a methodology for stress-based reliability analysis taking account of the scatter. The effects of reliability and the confidence level on fracture growth were considered, and the fatigue life with the confidence level and the reliability was given by using the one-sided allowance factor statistics analysis principle. The fatigue data model was optimized under the laser load by using the fracture mechanics theory. The one-sided allowance coefficient was revised, which was closer to the lower limit of the confidence of the matrix hundred rank value, and the usability and validity of the revision model were confirmed by analyzing the tentative data of 7050 aluminum alloy after LSP.

4.4.2 Analysis of Statistics Model

Usually, there should be enough test specimens so that the security fatigue life and the fatigue strength obtained from the fatigue experiment are requested to a certain confidence level. But to the actual components, there are only a few test specimens (5–11 numbers) as the result of the economic problems. If the number of test specimens cannot satisfy the request of the observed value, fatigue life with the confidence level γ and reliability p could be given by using the one-sided allowance factor k [86].

As the logarithm fatigue life follows the normal distribution, the logarithm security fatigue life x_p with the reliability p could be calculated as

$$x_p = \mu + u_p \sigma \tag{4.43}$$

where to any assigned reliability level p, the value of u_p may be determined, which makes some individual logarithm fatigue life with p in the matrix become bigger than x_p. While as to the random variable function ε, the probability of the security life is smaller than the matrix's true value, which is expressed as the confidence level γ [86],

$$P(\bar{x} + k\beta s < \mu + u_p \sigma) = \gamma \tag{4.44}$$

As assumed that ε follows the normal distribution N, formula (4.43) could be changed into as follows:

$$\mu + u_p\sigma = E(\varepsilon) + u_\gamma\sqrt{\text{Var}(\varepsilon)}. \tag{4.45}$$

According to the statistics theory, it could be defined by

$$E(\varepsilon) = E[\bar{x} + k\beta s] = E(\bar{x}) + kE(\beta s) = \mu + k\sigma \tag{4.46}$$

$$\text{Var}(\varepsilon) = \text{Var}(\bar{x}) + k^2\beta^2\text{Var}(s)$$

$$= \frac{\sigma^2}{n} + k^2\beta^2\frac{\sigma^2}{n-1}\left\{ n - 1 - 2\left[\frac{\Gamma\left(\frac{n}{2}\right)}{\Gamma\left(\frac{n-1}{2}\right)}\right]^2 \right\}$$

$$= \frac{\sigma^2}{n} + k^2\left\{ \frac{(n-1)\sigma^2}{2}\left[\frac{\Gamma\left(\frac{n-1}{2}\right)}{\Gamma\left(\frac{n}{2}\right)}\right]^2 - \sigma^2 \right\} \tag{4.47}$$

As $n = 5$, the formula (4.47) could be written as

$$\text{Var}(\varepsilon) = \sigma^2\left[\frac{1}{n} + \frac{k^2}{2(n-1)}\right] \tag{4.48}$$

So, the formula (4.43) could be acquired as

$$\mu + u_p\sigma = \mu + k\sigma + u_\gamma\sigma\sqrt{\frac{1}{n} + \frac{k^2}{2(n-1)}} \tag{4.49a}$$

$$k = \frac{u_p - u_\gamma\sqrt{\frac{1}{n}\left[1 - \frac{u_\gamma^2}{2(n-1)}\right] + \frac{u_p^2}{2(n-1)}}}{1 - \frac{u_\gamma^2}{2(n-1)}} \tag{4.49b}$$

where k is the one-sided allowance factor, n is the number of the observed values, u_p and u_γ are the partial volumes of normal related with the reliability and confidence, respectively; the values of u_γ under the confidence levels of 90 and 95 % are 1.282 and 1.645, respectively [86]. The security logarithm life with the confidence level γ and reliability p could be expressed as

$$\hat{x}_p = \bar{x} + k\beta s \tag{4.50}$$

where u_p, u_γ, and n are all known, the factor k could be obtained by the formula (4.49b), and then security life with the confidence level γ and the reliability p could be obtained by the formula (4.50) as

$$\hat{N_P} = \lg^{-1} \hat{x_p} \tag{4.51}$$

That is to say, some individual fatigue life with p at the confidence level of γ is bigger than $\hat{N_P}$ in the matrix at least.

4.4.3 LSP and Fatigue Experiment

LSP is a fatigue enhancement surface treatment for metallic materials in which residual compressive stresses are mechanically produced into the surface [87–89]. LSP in a confined geometry was distinctly different from direct ablation because the coating layer was vaporized in the case of confined geometry, whereas the work-piece itself was vaporized in the case of direct ablation. The collisions among the vaporized particles in the vaporized layer must be considered in the case of confined geometry, and the recoil and plasma pressures were of interests to calculate the pressure on the substrate surface, as shown in Fig. 4.21. 7050-T7451 aluminum alloy is used as the test specimen, which is prefabricated with cracks on its surface. The sizes of samples are shown in Fig. 4.22. The material parameters are shown in tables. The fatigue test is completed on PLG-100C high-frequency fatigue testing machine. The stress fatigue lives of the test samples before and after LSP were compared. Fatigue experiments with the stress level from 60 to 140 MPa of the test samples after the processing were carried out (Table 4.5).

The experimental results showed that stress fatigue life of the test sample after LSP is 1–2 times of that of the non-shocked test sample. With regard to the non-through crack surface after LSP, the residual stress may witness significant variation

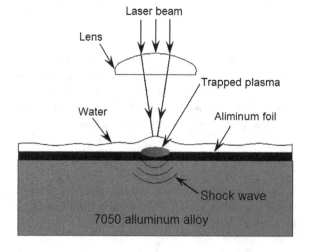

Fig. 4.21 Laser-shock peening on 7050 aluminum alloy. Reprinted from Ref. [90], Copyright 2012, with permission from Elsevier

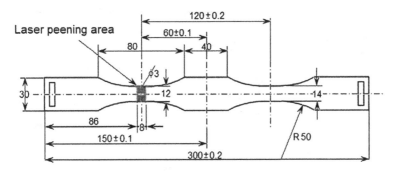

Fig. 4.22 Test specimen. Reprinted from Ref. [90], Copyright 2012, with permission from Elsevier

Table 4.5 7050-T7451 mechanical properties

Trade mark	Supply condition	Thickness mm	Sampling direction	Tensile strength σ_b (MPa)	Non-proportional extension strength $\sigma_{0.2}$ (MPa)	Elongation after break δ_S (%)
				No less than		
7050	T7451	60–76	L	503	434	9
			LT	503	434	8
			ST	469	407	3
		>76–102	L	496	427	9
			LT	496	427	6
			ST	469	400	3

Notes L is portrait, *LT* is landscape orientation, *ST* is thickness

under the instant LSP, which may further incur the variation of crack surface stress intensity factor (SIF) along the crack front.

The experimental result was carried on data statistics processing, the stress-crack initiation life data are shown in Fig. 4.23. Figure 4.24 gives the fitting effect of the one-sided allowance factor method to the tentative data with the conventional reliability of 0.95 and different confidence levels; it could be known from Fig. 4.24 that the fatigue reliability assessment result tends security along with increasing the confidence level.

4.4.4 Revision of Statistics Model

The one-sided allowance coefficient is used to determine the lower limit of the matrix hundred rank values confidence level, but the absolute value of k is often very big. Therefore, the confidence interval is wide and the obtained difference between the lower limit confidence level and the matrix's true one is significant.

Fig. 4.23 S-N data for 7050
aluminum alloy. Reprinted
from Ref. [90], Copyright
2012, with permission from
Elsevier

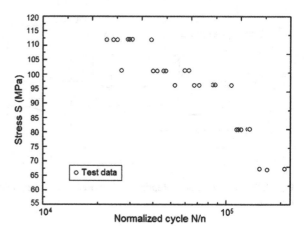

Fig. 4.24 Typical S–N
curves of 7050 aluminum
alloy for 0.95 reliability with
different confidence.
Reprinted from Ref. [90],
Copyright 2012, with
permission from Elsevier

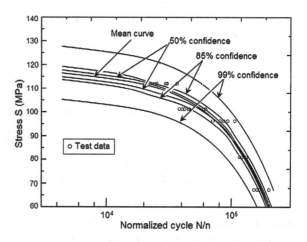

So, the concept of the one-sided allowance coefficient is needed to give lower
confidence of matrix hundred rank values with the confidence level which saves the
test sample with the same precision. Regarding to the random variable, the concept
could be given as

$$\eta = \overline{X} + u_R \beta s \tag{4.52}$$

where η obeys the normal distribution approximately. Its average value and the
variance are calculated as follows, separately:

$$E(\eta) = E(\overline{X} + u_p \beta s) = E(\overline{X}) + u_p E(\beta s) = \mu + u_p \sigma \tag{4.53a}$$

$$D(\eta) = D(\overline{X} + u_p \beta s) = D(\overline{X}) + u_p \beta^2 D(s) \tag{4.53b}$$

So, $D(\overline{X})$ could be calculated as

$$D(\overline{X}) = \sigma^2/n \tag{4.54}$$

From variable of the freedom degree, it could be obtained as

$$(n-1)s^2/\sigma^2 = \chi^2 \tag{4.55a}$$

$$s = \frac{\chi\sigma}{\sqrt{n-1}} \tag{4.55b}$$

So $D(s)$ could be calculated as

$$D(s) = \frac{\sigma^2}{n-1}D(\chi) \tag{4.56}$$

The character of the variable of χ^2 could be known as

$$E(\chi^2) = n-1 \tag{4.57a}$$

$$E(\chi) = \sqrt{2}\,\frac{\Gamma\left(\frac{n}{2}\right)}{\Gamma\left(\frac{n-1}{2}\right)} \tag{4.57b}$$

So that $D(s)$ could be expressed as

$$D(s) = \sigma^2\left\{1 - \frac{2}{n-1}\left[\frac{\Gamma\left(\frac{n}{2}\right)}{\Gamma\left(\frac{n-1}{2}\right)}\right]^2\right\} = \sigma^2\left(1 - \frac{1}{\beta^2}\right) \tag{4.58a}$$

$$D[\eta] = \frac{\sigma^2}{n} + u_p^2\left(\beta^2 - 1\right) \tag{4.58b}$$

And the standard normal variable could be obtained as

$$U = \frac{\eta - E(\eta)}{\sqrt{D(\eta)}} = \frac{\left(\overline{X} + u_p\beta s\right) - \left(\mu + u_p\sigma\right)}{\sigma\sqrt{\frac{1}{n} + u_p^2\left(\beta^2 - 1\right)}} \tag{4.59}$$

It could be proved that μ and χ^2 are independent from each other, thus the variable t could be obtained as [90]

$$t = \frac{u}{\sqrt{\frac{\chi^2}{\nu}}} = \frac{\left(\overline{X} + u_p\beta s\right) - \left(\mu + u_p\sigma\right)}{s\sqrt{\frac{1}{n} + u_p^2\left(\beta^2 - 1\right)}} \tag{4.60}$$

where t_y is supposed to be the t distribution hundred rank values, which means that

$$P(t_y \leq t) = \gamma \tag{4.61a}$$

$$P\left\{ \frac{\overline{X} + u_p \beta s - (\mu + u_p \sigma)}{s\sqrt{\frac{1}{n} + u_p^2(\beta^2 - 1)}} \leq t_r \right\} \tag{4.61b}$$

$$P\left\{ (\overline{X} + u_p \beta s) - (\mu + u_p \sigma) \leq t_y s \sqrt{\frac{1}{n} + u_p^2(\beta^2 - 1)} \right\} = \gamma \tag{4.61c}$$

$$P\left\{ (\overline{X} + u_p \beta s) - t_y s \sqrt{\frac{1}{n} + u_p^2(\beta^2 - 1)} \leq (\mu + u_p \sigma) \right\} = \gamma \tag{4.61d}$$

Thus, the new one-sided allowance coefficient could be defined as

$$m = u_p \beta - t_y \sqrt{\frac{1}{n} + u_p^2(\beta^2 - 1)} \tag{4.62}$$

And then γ can be defined as

$$P(\overline{X} + ms \leq \mu + u_p \sigma) = \gamma \tag{4.63}$$

So formula (4.56) could be wrote in the revision sample as

$$\hat{x}_p = \overline{X} + ms \tag{4.64}$$

That is to say that there is at least P with the confidence level of γ of the individuals x are bigger than $\hat{x}_p$ in the statistics samples.

Figure 4.25 gives the analysis chart of the ratio of the new and old allowance coefficients under the different confidence levels with the reliability of 0.95. It could be found that when the reliability and number of the test samples are determined, the higher the confidence level is, the more obvious the new one-sided allowance coefficient's superiority is. Figure 4.26 is the analysis chart of the ratio of allowance coefficients under different reliabilities with confidence level of 0.95. When the confidence level is 95 %, the data show that when the reliability P is big, its allowance coefficient ratio is also big. That is to say, when the confidence level and the number of the test samples are determined, the higher the reliability is, the more obvious the new one-sided allowance coefficient's superiority is. The fitting effect of the new and old allowance coefficients to the P–S–N tentative data with the confidence level of 0.95 is shown in Fig. 4.27. It could be clearly seen that the confidence level's lower limit obtained by using the new one-sided allowance coefficient is closer to the matrix hundred rank value, and the new one-sided

Fig. 4.25 Typical allowance coefficients of 7050 aluminum alloy for 0.95 reliability with different reliabilities (β = const.: k = allowance factor; m = new allowance coefficient.). Reprinted from Ref. [90], Copyright 2012, with permission from Elsevier

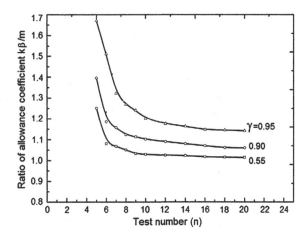

Fig. 4.26 Typical allowance coefficients of 7050 aluminum alloy for 0.95 reliability with different reliabilities (β = const.: k = allowance factor; m = new allowance coefficient.). Reprinted from Ref. [90], Copyright 2012, with permission from Elsevier

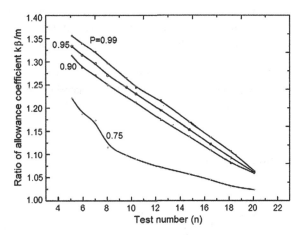

Fig. 4.27 Typical S–N curves of 7050 aluminum alloy for 0.95 confidence with new (m) and old $k(\beta)$ allowance coefficients. Reprinted from Ref. [90], Copyright 2012, with permission from Elsevier

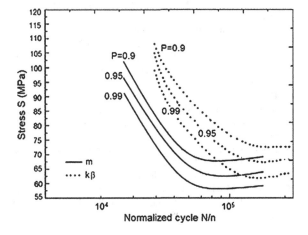

allowance coefficient saves more test samples in the same situation of precision. By comparing the 7050 aluminum alloy tentative data before and after the LSP, it demonstrates that this method gives the fatigue life of the statistical data. The revision coefficient would also save the experimental work load in the experiment, i.e., the less of the test number, the more obvious the revised coefficient's superiority is. The bigger of the reliability and confidence level are, the more obvious superiority of the revised one-sided allowance coefficients are.

4.4.5 Conclusions

One-sided allowance factor statistical analysis method was proposed, and the fatigue life with the confidence level and the reliability has been given in this chapter.

The lower limit of the confidence level obtained with the new one-sided allowance coefficient is closer to the matrix hundred rank values. In addition, the new one-sided allowance coefficient saves much more test samples. Thus, the one-sided allowance coefficient curve statistical model has reflected the function of the data quantity in the fatigue reliability assessment.

Hence, further work should be carried out to perform a more accurate relation on stable crack growth for elastic–plastic materials and the fatigue life with the confidence level and the reliability.

References

1. Montross Charles S et al (2002) Laser shock processing and its effects on microstructure and properties of metal alloys: a review. Int J Fatigue 24:1021
2. White RM (1963) Elastic wave generation by electron bombardment or electromagnetic wave absorption. J Appl Phys 34:2123–2124
3. Hu Y, Yao Z (2006) 3-D FEM simulation of laser shock processing. Surf Coat Technol 201:1426–1435
4. Peyre P et al (1996) Laser shock processing of aluminium alloys, application to high cycle fatigue behaviour. Mater Sci Eng A 210:102
5. Ocaña JL et al (2004) Experimental assessment of the influence of irradiation parameters on surface deformation and residual stresses in laser shock processed metallic alloys. Appl Surf Sci 238:501
6. El-Dasher Bassem S et al (2006) Surface deformation behavior of beta solution treated and overaged Ti-6Al-4 V during laser shock processing. J Appl Phys 99:103–506
7. Alfredsson B, Olsson M (1999) Standing contact fatigue. Fatigue Fract Eng Mater Struct 22 (3):225–237
8. MackAldener M, Olsson M (2000) Interior fatigue fracture of gear teeth. Fatigue Fract Eng Mater Struct 234:283–292
9. Tada H, Paris PC (2001) The stress analysis of cracks handbook. ASME Press, New York
10. Skorupa M (1999) Load interaction effects during fatigue crack growth under variable amplitude loading-a literature review Part II: qualitative interpretation. Fatigue Fract Eng Mater Struct 22(10):905–926

11. Skorupa M et al (1999) Fatigue crack growth in a structural steel under single and multiple periodic overload cycles. Fatigue Fract Eng Mater Struct 22(10):879–887
12. Hatamleh O et al (2007) Int J Fatigue 29:421
13. Rankin Jon E et al (2003) The effects of process variations on residual stress in laser peened 7049 T73 aluminum alloy. Mater Sci Eng, A 349:279
14. Nalla RK et al (2003) On the influence of mechanical surface treatments/deep rolling and laser shock peening/on the fatigue behavior of Ti/6Al/4V at ambient and elevated temperatures. Mater Sci Eng, A 355:216
15. Masaki K et al (2007) Effects of laser peening treatment on high cycle fatigue properties of degassing-processed cast aluminum alloy. Mater Sci Eng, A 468:171
16. Sano Yuji et al (2006) Retardation of crack initiation and growth in austenitic stainless steels by laser peening without protective coating. Mater Sci Eng, A 417:334
17. Sadananda K, Vasudevan AK (2005) Fatigue crack growth behavior of titanium alloys. Int J Fatigue 27:1255
18. Sánchez-Santana U et al (2006) Wear and fricti of 6061-T6 aluminum alloy treated by laser shock processing. Wear 260:847
19. Ren XD et al (2006) Study of the effect of coatings on mechanical properties of TC4 titanium alloy during laser shock processing. Mater Sci Forum 532:73
20. Zhang Y et al (2006) Study of the mechanism of overlays acting on laser shock waves. J Appl Phys 100:103517
21. Montross Charles S et al (2008) Laser shock processing and its effects on microstructure and properties of metal alloys: a review. Int J Fatigue 24:1021–1027
22. White RM (1963) Elastic wave generation by electron bombardment or electromagnetic wave absorption. J Appl Phys 34:2123–2124
23. Ocaña JL et al (2004) Experimental assessment of the influence of irradiation parameters on surface deformation and residual stresses in laser shock processed metallic alloys. Appl Surf Sci 238:501–509
24. El-Dasher Bassem S et al (2006) Surface deformation behavior of beta solution treated and overaged Ti-6Al-4V during laser shock processing. J Appl Phys 99:103506–103517
25. Peyre P et al (1996) Laser shock processing of aluminium alloys, Application to high cycle fatigue behaviour. Mater Sci Eng, A 210:102–110
26. Alfredsson B, Olsson M (1999) Standing contact fatigue. Fatigue Fract Eng Mater Struct 3:225–237
27. MackAldener M, Olsson M (2000) Interior fatigue fracture of gear teeth. Fatigue Fract Eng Mater Struct 234:283–292
28. Tada H, Paris PC (2001) The stress analysis of cracks handbook. ASME Press, New York, pp 54–59
29. Skorupa M (1999) Load interaction effects during fatigue crack growth under variable amplitude loading-a literature review: Part II. Qualitative interpretation Fatigue Fract Eng Mater Struct 22(10):905–926
30. Qu J, Wang X (2006) Solutions of T-stresses for quarter-elliptical corner cracks in finite thickness plates subject to tension and bending. Int J Press Vessel Pip 83:593–606
31. Hills DA et al (1996) Solution of crack problems: the distributed dislocation technique. Kluwer, Dordrecht
32. Yuji Sano et al (2006) Retardation of crack initiation and growth in austenitic stainless steels by laser peening without protective coating. Mater Sci Eng, A 417:334–345
33. Khraishi TA, Demir I (2003) On Cauchy singular integrals and stress intensity factors for 2D mode I crack in an infinite solid. Mech Res Commun 30:353–364
34. Fujimoto WT (1976) Determination of crack growth and fracture toughness parameters for surface flaws emanating from fastener holes. In: Proceedings of the AIAA/ASME/SEA 17th structures, structural dynamics and material conference, King of Prussia (PA)
35. Kim JH, Lee SB (2000) Prediction of crack opening stress for part-through cracks and its verification using a modified strip-yield model. Eng Fract Mech 66:1–14

36. Sadananda K, Vasudevan AK (2005) Fatigue crack growth behavior of titanium alloys. Int J Fatigue 27:1255–1267
37. Le Delliou P, Barthelet B (2007) New stress intensity factor solutions for an elliptical crack in a plate. Nucl Eng Des 237:1395–1405
38. DeWald Adrian T et al (2004) Assessment of tensile residual stress mitigation in alloy 22 welds due to laser peening. J Eng Mater Technol 126:465
39. Ballard P et al (1991) J Phys IV 1:487–494
40. Wu XR (1992) Analytical wide-range weight functions for various finite cracked bodies. Eng Anal Bound Elem 9:307
41. Ren XD et al (2009) Investigation of the stress intensity factor changing on the hole crack subject to laser shock processing. Mater Des 30:2769–2773
42. Zhao W, Wu XR (1990) Stress-intensity factor evaluation by weight function for surface crack in edge notch. Theor Appl Fract Mech 13:225
43. Zhao W et al (1989) Weight function method for three dimensional crack problems I: Basic formulation and application to an embedded elliptical crack in finite plates. Eng Fract Mech 34:593–607
44. Shen G, Glinkag G (1991) Determination of weight functions from reference stress intensity factors. Theor Appl Fract Mech 15:247
45. Zhao W et al (1989) Weight function method for three dimensional crack problems II. Eng Fract Mech 34:324–609
46. Chu J et al (1995) Effects of laser-shock processing on the microstructure and surface mechanical properties of Hadfield manganese steel. Metall Mater Trans A 26A(6):1507–1517
47. Ren XD et al (2009) Influence of compressive stress on stress intensity factor of hole-edge crack by high strain rate laser shock processing. Mater Des 30:3512–3517
48. Zhao W, Wu XR (1990) Stress-intensity factor evaluation by weight function for surface crack in edge notch. Theor Appl Fract Mech 13:225–241
49. Zhao W et al (1989) Weight function method for three dimensional crack problems: I. Basic formulation and application to an embedded elliptical crack in finite plates. Eng Fract Mech 34:593–607
50. Zhao W et al (1989) Weight function method for three dimensional crack problems: II. Eng Fract Mech 34:324–609
51. Daniewicz SR (1998) A modified strip yield model for prediction of plasticity induced closure in surface flaws. Fatigue Fract Eng Mater Struct 21:885–901
52. Rubio-González C et al (2004) Effect of laser shock processing on fatigue crack growth and fracture toughness of 6061-T6 aluminum alloy. Mater Sci Eng, A 386:291–295
53. Ren XD et al (2011) Mechanical properties and residual stresses changing on 00Cr12 alloy by nanoseconds laser shock processing at high temperatures. Mater Sci Eng, A 528:1949–1953
54. Yang JM et al (2001) Laser shock peening on fatigue behavior of 2024-T3 Al alloy with fastener holes and stopholes. Mater Sci Eng, A 298:296–299
55. King A et al (2006) Effects of fatigue and fretting on residual stresses introduced by laser shock peening. Mater Sci Eng, A 435–436:12–18
56. Ren XD et al (2010) Effects of laser shock processing on 00Cr12 mechanical properties in the temperature range from 25 °C to 600 C. Appl Surf Sci 257:1712–1715
57. Majzoobi GH et al (2009) The effects of deep rolling and shot peening on fretting fatigue resistance of Aluminum-7075-T6. Mater Sci Eng, A 516:235–247
58. De PS et al (2009) Effect of microstructure on fatigue life and fracture morphology in an aluminum alloy. Scripta Mater 60:500–503
59. Sanchez-Santana U et al (2006) Wear and friction of 6061-T6 aluminum alloy treated by laser shock processing. Wear 2006(260):847–854
60. Harold Luong, Hill Michael R (2010) The effects of laser peening and shot peening on high cycle fatigue in 7050-T7451 aluminum alloy. Mater Sci Eng, A 527:699–707
61. Kiyotaka Masaki et al (2007) Effects of laser peening treatment on high cycle fatigue properties of degassing-processed cast aluminum alloy. Mater Sci Eng 468:171–175

62. Clauer AH et al (1983) The effects of laser shock processing on the fatigue properties of -T3 aluminum. In: Metzbower EA (ed) Lasers in materials processing. American Society for Metals, Metals Park, pp 7–22
63. Ren XD et al (2009) Influence of compressive stress on stress intensity factor of hole edge crack by high strain rate laser shock processing. Mater Des 30:3512–3517
64. Zhang YK et al (2009) Investigation of the stress intensity factor changing on the hole crack subject to laser shock processing. Mater Des 30:2769–2773
65. Sohel Rana M et al (2009) The effect of hole shape on the extent of fatigue life improvement by cold expansions. Eng Fail Anal 16:2081–2090
66. Ren XD et al (2013) The effects of residual stress on fatigue behavior and crack propagation from laser shock processing-worked hole. Mater Des 44:149–154
67. SAE AMS 2546 (2004) http://www.sae.org. Accessed 17 Mar 2014
68. de Matos PFP et al (2005) Reconstitution of fatigue crack growth in Al-alloy 2024-T3 open-hole specimens using photomicrography techniques. Eng Fract Mech 72:2232–2246
69. Barlow KW, Chandra R (2005) Fatigue crack propagation simulation in an aircraft engine fan blade attachment. Int J Fatigue 27:1661–1668
70. Sadananda K, Vasudevan AK (2005) Fatigue crack growth behavior of titanium alloys. Int J Fatigue 27:1255–1266
71. Zhao YX et al (2009) Interaction and evolution of short fatigue cracks. Fatigue Fract Eng Mater Struct 22(66):459–467
72. Schutz D, Heuler P (1994) The significance of variable amplitude fatigue testing. In: Amzallag C (ed) Automation in fatigue and fracture; testing and analysis, ASTM STP1231, pp 201–220
73. Niemand NJ, Wonnenbury J (1997) Vehicle durability programmes. J Eng Int Soc 7:22–26
74. Lewis C (1999) Simulating the rocky road with fatigue testing. Mater World 7(3):139–140
75. Nisitani H, Chen DH (1984) Stress intensity factor for a semi-elliptical surface crack in a shaft under tension. Trans Jpn Soc Mech Eng 50:1077–1082
76. Zhao YG, Ono T (2001) Moment for structural reliability. Struct Saf 23:47–75
77. Weibull W (1939) A statistical theory of the strength of materials. R Swed Inst Eng Res 151
78. Bomas H et al (1999) Application of a weakest-link concept to the fatigue limit of the bearing steel SAE 52100 in abainitic condition. Fatigue Fract Eng Mater Struct 22(9):733–741
79. Hild F, Roux S (1991) Fatigue initiation in heterogeneous brittle materials. Fatigue Fract Eng Mater Struct 18:409–414
80. Chantier I et al (2000) A probabilistic approach to predict the very high cycle fatigue behaviour of spheroidal graphite cast iron structures. Fatigue Fract Eng Mater Struct 23:173
81. Morel F, Flacelie're L (2005) Data scatter in multi-axial fatigue: from the infinite to the finite fatigue life regime. Int J Fatigue 27(9):1089–1101
82. Nadot Y, Denier V (2004) Fatigue failure of suspension arm: experimental analysis and multiaxial criterion. Eng Fail Anal 11:489–499
83. Bjerager P (1990) On computation methods for structural reliability analysis. Struct Saf 9:79–96
84. Tovo R (2001) On the fatigue reliability evaluation of structural components under service loading. Int J Fatigue 23:587–598
85. You BR, Lee SB (1996) A critical review on multiaxial fatigue assessments of metals. Int J Fatigue 18(4):235–244
86. Zhentong Gao et al (1999). Fatigue properties of experimental design and data processing. Beijing University of Aeronautics and Astronautics Press, Beijing, pp 19–21
87. Montross CS et al (2002) Laser shock processing and its effects on microstructure and properties of metal alloys: a review. Int J Fatigue 24:1021
88. Peyre P et al (1996) Laser shock processing of aluminium alloys, application to high cycle fatigue behaviour. Mater Sci Eng, A 210:102
89. Masaki K et al (2007) Effects of laser peening treatment on high cycle fatigue properties of degassing-processed cast aluminum alloy. Mater Sci Eng, A 468:171
90. Ren XD et al (2012) A model for reliability and confidence level in fatigue statistical calculation. Theor Appl Fracture Mech

Chapter 5
Conversion Model of Graphite

Abstract This chapter gives a well-rounded presentation of the continuous synthesis of UNCD via laser shock processing (LSP) of graphite particles suspended in water by a Nd:YAG laser system with high power density (10^9 W/cm^2) and short pulse width at room temperature and normal pressure, which yielded the ultra-nano-crystalline diamond in size of about 5 nm. X-ray diffraction, high-resolution transmission electron microscopy, and laser Raman spectroscopy were used to characterize the nano-crystals. The method studied is helpful in understanding the formation mechanism and enhancing the yield rate of nano-diamond.

5.1 Introduction

Carbon has attracted much attention for its various allotropes with different physical and chemical natures originated from the variety of *s-p* orbital hybridization [1], such as graphite, diamond, fullerenes, and carbon nanotubes. The phase transformations among these forms are a meaningful subject especially the formation of diamonds. Diamond has been obtained by many methods, including high-temperature and high-pressure conditions [2], detonation [3], and chemical vapor deposition [4]. Recently, the method of pulsed laser ablation of carbon targets in liquid (PLAL) has been intensively studied. Yang [5] reported the preparation of nano-diamond by pulsed laser ablation on the graphite target. The size of nano-diamond obtained is about 40–300 nm. However, the synthesis is batch-type and relatively low efficiency by ejecting from the targets. And the mechanism of the PLAL method remains unclear because it is difficult to precisely measure the synthesis condition. Sun et al. [6] converted graphite into ultra-nano-crystalline diamond (UNCD) via laser irradiation of suspension of graphite particles by using a lower laser power density (10^6 W/cm^2) with long-pulsed (1.2 ms) laser. Therefore, the transformation mechanism even the product obtained is different for the different experimental methods and parameters (laser power density, pulse width). In addition, the higher laser power density is more favorable for carbon phase

© Springer-Verlag Berlin Heidelberg 2015
X. Ren, *Laser Shocking Nano-Crystallization and High-Temperature Modification Technology*, DOI 10.1007/978-3-662-46444-1_5

transformation [7]. The UNCD owns outstanding mechanical, electronic transport, chemical and biocompatibility properties [8, 9]. In this paper, we describe that the continuous synthesis of UNCD via laser shock processing (LSP) of graphite particles suspended in water by a Nd:YAG laser system with high power density (10^9 W/cm^2) and short pulse width at room temperature and normal pressure. The method combines the advantages of standard PLAL and chemical methods. As compared to the conventional nano-diamond, the average size of UNCD is about 5 nm. Therefore, it is necessary to modify the parameters of the used laser to obtain different products, which are helpful in understanding the formation mechanism and enhancing the yield rate of nano-diamond.

5.2 A Conversion Model of Graphite to Ultra-nano-crystalline Diamond via Laser Processing at Ambient Temperature and Normal Pressure

The synthesis mechanism of ultra-nano-crystalline diamond via LSP of graphite suspension was presented at room temperature and normal pressure, which yielded the ultra-nano-crystalline diamond in size of about 5 nm. X-ray diffraction, high-resolution transmission electron microscopy, and laser Raman spectroscopy were used to characterize the nano-crystals. The transformation model and growth restriction mechanism of high power density with short-pulsed laser shocking of graphite particles in liquid were put forward.

5.2.1 Experiment and Method

Graphite particles (purity of 99.99 %) were mixed with pure water in a cylindrical vessel, and the final concentration of graphite suspension was 0.02 g/ml. The graphite suspension was dispersed for about an hour in supersonic cleaner before laser irradiation. The experimental setup is shown in Fig. 5.1, which was performed by using a Nd:YAG laser (10^9 W/cm^2, 1064 nm) with a pulse width of 10 ns.

The Nd:YAG laser energy was maintained at 500 mJ with a repetition rate of 5 Hz in order to prevent splashes of liquid from reaching the prism or lens at higher energies. In order to fully absorb the laser energy, the beam was focused into a spot-size diameter of approximately 1 mm under the surface of water of 2–3 mm. The graphite suspension was stirred using a magnetic stirrer in the experimental process, and none of the ripple could be observed at the surface of the graphite suspension. After two hours of LSP, the acquisition was oxidized and purified by perchloric acid (HClO$_4$) until the liquid turned to be slightly colored (usually gray) due to the suspension of nanoparticles. Then, the purified particles were added into a vacuum drying oven to get dry.

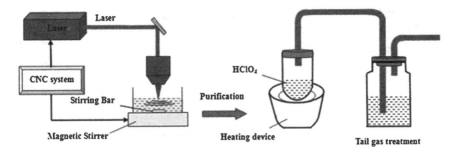

Fig. 5.1 Schematic of the LSP and purification method. Reprinted from Ref. [10], Copyright 2014, with permission from Elsevier

X-ray diffraction (XRD), Raman spectroscopy, and high-resolution transmission electron microscopy (HRTEM) were performed to characterize the products after LSP. The purified products with alcohol were dispersed for 9 min and then pipetted onto a carbon-coated TEM grid. XRD was performed on a D8 advance diffractometer. Laser Raman spectroscope (DXR, excitation wavelength 532 nm) was utilized to study the bonding configuration. High-resolution TEM (JEM-2100 microscope, 200 kV) was employed to identify the compositional, structural, and morphological information of the products.

5.2.2 Results

The detections of XRD, HRTEM, and Raman spectroscopy were implemented to identify the transformation of graphite particles into UNCD. Figure 5.2 presents the XRD spectra of the raw graphite particles, products before and after purification by perchloric acid ($HClO_4$) with LSP. Some obvious changes have occurred in these XRD spectra. Many impure peaks appear in the sample without purification after laser treatment. Therefore, it is necessary to purify the products by perchloric acid for the sake of removing non-diamond carbon phase and other impurities. The raw material has a high degree of ordering and good crystalline with both the sharp diffraction peak and the small full width at half maximum. Moreover, the intensity of the peaks is evidently subdued and even disappeared after treatment, which illustrates that the crystallinity and ordering degree of the particles decrease after LSP. What's more important is that the new X-ray diffraction peaks at the angles $2\theta = 43.9$ °C and 75.3 °C are well corresponding to (111) and (220) reflections of diamond lattice. The observed peak broadening at $2\theta = 55$ °C for purified products is attributed to the still existence of small graphite powders which have been oxidized into nano-scale by strong acid.

The diamond formation is also confirmed by Raman spectroscopy and HRTEM observation. Figure 5.3a shows the Raman spectrum in the region between 500 and 2000 cm^{-1} of the sample. It is clear that the nano-diamond is proven by a sharp,

Fig. 5.2 The XRD spectra of the raw graphite particles, products with and without purification. Reprinted from Ref. [10], Copyright 2014, with permission from Elsevier

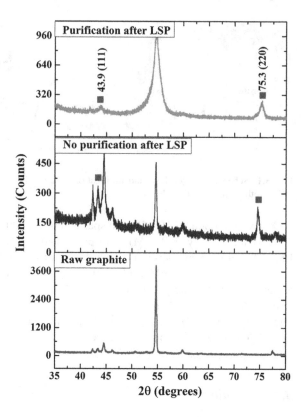

intense peak at 1327.7 cm^{-1} indicating sp^3 bonding structure of nano-diamond. The downward shift of the peak from its value of 1332 cm^{-1} for bulk diamond is consistent with the phonon confinement model [11] for particles less than 5 nm in size, and the width of Raman peak generally appears in the nanoparticles. Moreover, Obraztsova et al. [12] also reported that the Raman characteristic peaks of diamond would shift toward lower or higher wave number, if there is tensile stress or compressive stress. Therefore, there exists tensile stress in UNCD growth by LSP. The broaden peak at 1585 cm^{-1} is graphite peak with C–C sp^2 bond, and this is so-called the G-peak. The results show that the Raman scattering cross section of diamond is one-sixtieth of that graphite, and there are still graphite particle oxidized into nano-sizes after the acid pickling. Based on the phonon restriction model, the relationship between the half width of Raman peak and the particle size could be expressed as [13, 14],

$$\Gamma = 2.990 + 0.185/L \tag{5.1}$$

where Γ is the half width of Raman peak (μm), and L is the particle size (μm). According to the FWHM (42.84) of the diamond peak, the diamond size is obtained of about 5 nm.

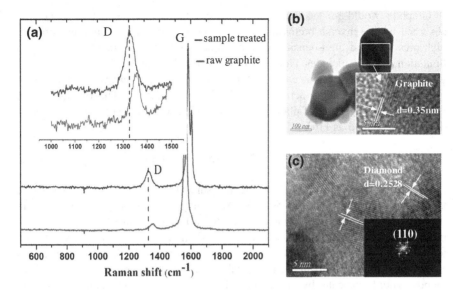

Fig. 5.3 The Raman spectroscopy and TEM observation of the formation diamond. **a** Typical Raman spectra. **b** The graphite particle HRTEM image. **c** HRTEM bright field image, and the diffraction image, ⟨110⟩ family plane of diamond. Reprinted from Ref. [10], Copyright 2014, with permission from Elsevier

The HRTEM image of raw graphite in Fig. 5.3b and the XRD spectra of the raw graphite particles in Fig. 5.2 demonstrate that the raw material is flake graphite. A typical HRTEM image of products after LSP is given in Fig. 5.3c, which shows that there are a lot of spherical microcrystalline in sizes of 4–6 nm. The atomic planes with the spacing of 0.2528 nm are assigned as the (110) plane with the diamond. This result confirms once again that UNCD has been obtained during LSP of graphite particles suspended in water.

5.2.3 Discussion

5.2.3.1 Phase Transition Mechanism

Graphite is easy to sublimate because that its sublimation temperature is less than the melting temperature. Savvatimskiy [15] reported that graphite has very high melting point of about 4800 nm and its assigned sublimation temperature is just about 3900 K. Therefore, graphite would gasify before melting in the laser heating. Based on the mechanism of interaction of laser and material [5], the laser energy is firstly absorbed by graphite particles and converted into heat, and finally forming the temperature field through thermal conduction.

Graphite would be gasified in this process firstly. As more laser energy is absorbed by the inverse bremsstrahlung and photoionization, a high-temperature, high-pressure, and high-density plasma plume would be generated for the liquid limitation, which includes carbon atom, carbon ion, electron, and radicals of graphite. The carbon plasma could be found as a particle cluster with high degree of ionization, high chemical activity, and high kinetic energy. The relationship of temperature and pressure of plasma plume as LSP of graphite particles in the liquid could be described as [16],

$$T = -0.1163P^2 + 15.66P + 4000 \tag{5.2}$$

According to the formula (5.2), the pressure (P) is about 15–20 GPa as the temperature (T) is about 4000–5000 K. According to phase diagram of carbon in Fig. 5.4, the carbon atoms in the plasma plume locate in liquid carbon area. The presence of transient liquid phase is the core of this model. As the decrease of the temperature and pressure for the inflation of plasma, these liquid carbon atoms would quickly enter into in diamond stable and graphite metastable region. Diamonds would nucleate by recrystallization in this thermodynamics condition. Simultaneously, particle cluster supplies abundant carbon source for diamond crystal nucleus. When the rarefaction wave following compression wave arrival, pressure reduced and a large degree of supercooling is obtained. The crystal nucleus

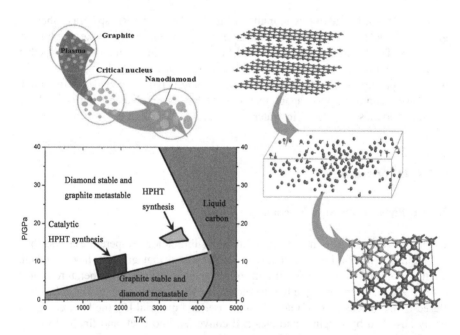

Fig. 5.4 Reaction mechanism, temperature and pressure phase diagram of carbon, and phase transition mechanism. Reprinted from Ref. [10], Copyright 2014, with permission from Elsevier

of diamond would grow in the process. However, the diamond growth process is accompanied by graphitization for which the pressure and the temperature reduce sharply. Therefore, we could say that high-energy LSP of graphite suspension is a solid–vapor–liquid–solid phase transformation process.

5.2.3.2   Growth Restriction Mechanism

Based on thermodynamics principle, the equilibrium phase boundary between graphite and diamond could be expressed by Wang et al. [17] and Wang and Yang [18]

$$\Delta p + p = 2.01 \times 10^6 T + 2.02 \times 10^9 (P_a) \tag{5.3}$$

where P and ΔP are the external pressure and the additional pressure induced by the surface stress, respectively, and T is the temperature. Based on the Laplace–Young equation [19], ΔP can be expressed as $2\gamma/r$. Therefore, the relationship between the sizes of nano-diamond and experimental conditions (temperature and pressure) is given by $r = 2\gamma/(2.01 \times 10^6 T + 2.02 \times 10^9 - P)$. This equation indicates that the temperature and pressure caused by the laser would influence the size of the diamond. In the range of temperature and pressure (4000–5000 K, 10–20 GPa) of plasma, the size of diamond is just few nanometers. However, we think that the final determination of diamond size is not only just the temperature and pressure, but also the growth velocity and growth time. Based on the Wison–Frenkel growth law, the relationship of the growth temperature T and the growth velocity V of the crystalline nucleus could be described as [20],

$$V = hf \exp(-E_a/RT)[1 - \exp(-\Delta G_m/RT)] \tag{5.4}$$

where h (0.206 nm) is the lattice constant of diamond nuclei, f ($2.506n^{13}$ Hz) is the thermal vibration frequency, E_a (2.4 is the T) is the molar adsorption energy of adatoms attached at the surface sites, R is the gas constant, and ΔG_m is the Gibbs free energy difference per mole which could be expressed as,

$$\Delta G_m = \Delta H_s \Delta T / T_v \tag{5.5}$$

where ΔH_s (355.80 kJ/mol) is the agglomerate enthalpy, T_v (about 4000 K) is the condensed temperature, ΔT is the degree of supercooling, and $\Delta T = T_v - T$.

According to Eqs. (5.4) and (5.5), the relationship of the growth velocity and the degree of supercooling could be expressed as,

$$V = hf \frac{\Delta H_s \Delta T}{RT_v(T_v - \Delta T)} \exp[-E_a/R(T_v - \Delta T)] \tag{5.6}$$

The short-pulsed laser with higher power density would impulse graphite particles to a vapor state under higher temperature. Meanwhile, the graphite particles suspension is easier to change heat. After laser shock pulse, the carbon plasma cools quickly which leads to a larger degree of supercooling ΔT. According to Eq. (5.6), the high degree of supercooling ΔT allows a rather high growth velocity of diamond nuclei. However, for a nanosecond pulse laser, the diamonds only grow in a very narrow range of temperature 4100–4004 K [21]. So the diamond crystal nucleus growth time is very short to the nanosecond laser. Meanwhile, the disordered structure formed at the diamond surface would restrict the epitaxial growth of the nano-diamonds. During the process of purification, the effect of strong oxidizer would also reduce the size of the diamond.

5.2.4 Conclusions

The method of LSP on the carbon material at room temperature and normal pressure has been studied. (I) The UNCD about 5 nm has been produced via LSP of graphite particles suspended in water. (II) Phase transition mechanism: high-energy with short-pulsed laser shocking of graphite particles is a solid–vapor–liquid–solid phase transformation process. (III) The nano-crystal diamond nucleates in the early stage of the plasma expansion and grows in subsequent rapid quenching. The high degree of supercooling and the short growth time dynamically determine the final size of diamonds.

References

1. Kitazawa S et al (2005) Formation of nanostructured solid-state carbon particles by laser ablation of graphite in isopropyl alcohol. Solids 66:555
2. Abbaschian R et al (2005) High pressure-high temperature growth of diamond crystals using split sphere apparatus. Diamond Relat Mater 14:1916
3. Chen PW et al (2000) Spherical nanometer-sized diamond obtained from detonation. Diamond Relat Mater 9:1722
4. Wang WL et al (2000) Nucleation and growth of diamond films on aluminum nitride by hot filament chemical vapor deposition. Diamond Relat Mater 9:1660
5. Yang GW (2007) Pulsed laser ablation and deposition of thin films. Prog Mater Sci 52:648
6. Sun J et al (2006) Distribution of interleukin-1 receptors in term human fetal membranes and decidua. J Mater Res 20:33
7. Gao N (2007) Effects of graphite nodules on crack growth behaviour of austempered ductile iron. Powder Metall Tchch 25:203
8. Shen L, Chen Z (2009) A numerical study of the imperfection effect on ultrananocrystalline diamond properties under different loading paths and temperatures. Compos Sci Technol 69:2075
9. Na C et al (2009) Advanced deposition characteristics of kinetic sprayed bronze/diamond composite by tailoring feedstock properties. Compos Sci Technol 69:463
10. Ren XD et al (2014) A conversion model of graphite to ultrananocrystalline diamond via laser processing at ambient temperature and normal pressure. Appl Phys Lett 105:021908
11. Prawer et al (2000) The raman spectrum of nanocrystalline diamond. Chem Phys Lett 332:93

12. Obraztsova ED et al (1998) Raman-spectroscopy of low-dimensional semiconductors. Carbon 36:821
13. Fauchet PM, Campbell IH (1988) Raman-spectroscopy of low-dimensional semiconductors. Crit Rev Solid State Mater Sci 14:S79
14. Richter H et al (1981) Measurements of the melting point of graphite and the properties of liquid carbon. Solid State Commun 39:625
15. Savvatimskiy AI (2005) Measurements of the melting point of graphite and the properties of liquid carbon (a review for 1963–2003). Carbon 43:1115
16. Wang JB, Yang GW (1999) Phase transformation between diamond and graphite in preparation of diamonds by pulsed-laser induced liquid-solid interface reaction. J Phys Condens Matter 11:7089
17. Wang CX et al (2005) Relaxorlike dielectric behavior in Ba0.7Sr0.3TiO3 thin films. J Appl Phys 97:066104
18. Wang CX, Yang GW (2005) Thermodynamics of metastable phase nucleation on nanoscale. Mater Sci Eng R 49:157
19. Sun J et al (2006) Ultrafine diamond synthesized by long-pulse-width laser. Appl Phys Lett 89:1831151
20. Wang CX et al (2005) Nucleation and growth kinetics of nanocrystals formed upon pulsed-laser ablation in liquid. Appl Phys Lett 87:201913
21. Tian F, Sun J (2009) Time amplifying techniques towards atomic time resolution. Chin J Lasers 1:3039

Printed in the United States
By Bookmasters